PREFACE

Particulate materials are substances consisting of individual particles, such as minerals, dusts, pollutants, catalysts, protective coatings, composites, toners, cosmetic powders, pharmaceutical drugs, fertilizers, cement, solids fuels, and foodstuff, which can be found in nature or manufactured through chemical and/or physical processes. They are the primary sources for particulate products, which account for over 60% of all the industrial products and are of significant importance to the global economy, society and environments. However, due to the diversity and intrinsic nature, manufacturing, handling and processing of particulate materials still face numerous challenges. Aiming at addressing these challenges, this book contains a selection of papers discussing the state-of-the-art research in particulate materials science that were presented at the UK-China Particle Technology Forum III held at Birmingham, UK on 3rd-6th July 2011.

The contents of this book are classified into four topics: synthesis, characterisation, processing and modelling. For synthesis of particulate materials, a range of advanced techniques are discussed, including spray-pyrolysis, supercritical fluid synthesis assisted with ultrasound, continuous synthesis using supercritical water, hydrothermal synthesis of nano-particulate materials and jet milling. In addition, how surface modification can be used to enhance the functionalities of particulate materials is illustrated. For characterisation, various methods for characterising particulate materials at both particle and system levels are introduced and how these properties affect the behaviour of particulate materials in various processes, such as inhalation, filling, and consolidation, is discussed. In the processing section, recent advances in particulate materials processing, such as capsule filling, micro-dosing, dry granulation, roll compaction, milling, fluidisation, particle separation, gas cleaning, agglomeration and pneumatic conveying, are presented. The last section concerns mathematical and numerical modelling in particulate materials, for which the book includes both analytical methods and advanced numerical methods, such as discrete element methods (DEM), computational fluid dynamics (CFD), lattice Boltzmann methods (LBM), coupled DEM/CFD and DEM/LBM, and their applications.

We would like to acknowledge all contributors for their efforts and support. We would like to express our sincere gratitude to the Engineering and Physical Science Researh Council (EPSRC), Royal Academy of Engineering, Birmingham Science City, Proctor and Gamble, AstraZeneca, Unilever and Freeman Technology for their financial support. We are grateful to the University of Birmingham, UK, the Institute of Process Engineering of Chinese Academy of Sciences, China, Particle Technology Subject Group of IChemE and Chinese Society of Particuology for their support and assistance in organising the UK-China Particle Technology Forum III that attracted more than 130 papers. We thank all members of the advisory, scientific and organising committees, and all reviewers for their advices, suggestions and guidance. Without their support, this book would not have been possible.

Chuan-Yu Wu and Wei Ge

September 2011

Contents

Processing

Modelling

Synthesis

PRODUCTION OF RESVERATROL NANOPARTICLES USING SOLUTION ENHANCED DISPERSION IN SUPERCRITICAL CO₂ WITH ENHANCED MASS TRANSFER BY ULTRASOUND

H.Y. Jin and Y.P. Zhao

School of Chemistry and Chemical Engineering, Shanghai   Jiaotong University, 800 Dongchuan RD. Shanghai   200240, China
Email: ypzhao@sjtu.edu.cn

1. INTRODUCTION

About 40% of the drugs or drug candidates are regarded as biopharmaceutic class II - low solubility and high permeability. Moreover, some side effects appear after the administration of such drugs in the gastrointestinal tract because of their toxicity, low bioavailability and poor absorption. To solve these problems, reducing the size of these drugs into micron or nanoparticles is a promising and effective way. The dissolution rate of these drugs can be enhanced due to increased surface area and chemical potential.[1] Many methods and techniques of reducing the size of ingredients in the pharmaceutical industry were established. Among them, the techniques based on supercritical fluids (SCF) as an antisolvent are considered the advanced approaches to obtain size controllable, tiny and free-solvent residual particles. These techniques consist of several main processes, such as supercritical antisolvent precipitation (SAS), aerosol solvent extraction system (ASES), and solution enhanced dispersion by supercritical fluid (SEDS) and supercritical antisolvent with enhanced mass transfer (SAS-EM). [2-5]

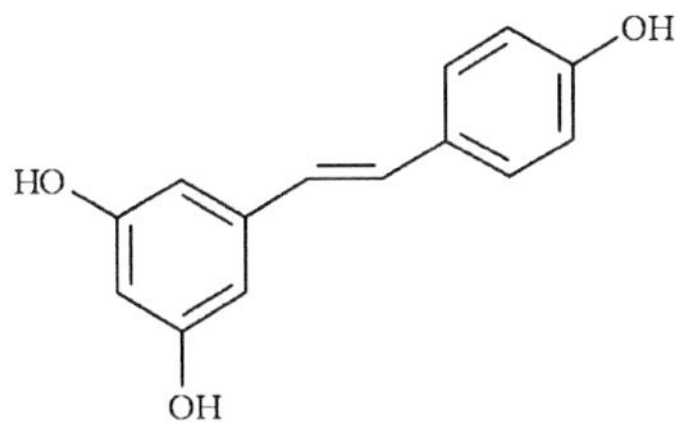

Figure 1 *The molecular structure of resveratrol*

Recently, a novel SEDS technique of solution enhanced dispersion with enhanced mass transfer by ultrasound in supercritical fluid, SEDS-EM, was introduced by the present authors. In this paper, resveratrol (3,5,4'-trihydroxy-trans-stilbene), a poorly water-soluble compound with promising applications in anti-cancer, anti-inflammatory, blood-sugar-lowering and cardioprotection, was selected as the modal drug candidate (The molecular structure is shown in Figure 1).[6-9] The modified supercritical antisolvent technique, solution enhanced dispersion in supercritical CO_2 with enhanced mass transfer (SEDS-EM), was proposed for the production of resveratrol nanoparticles.

2. MATERIALS AND METHOD

2.1 Materials

Resveratrol with a purity of 99.0% was obtained from Hangzhou Guang Lin Pharmaceutical Ltd (China). Dichloromethane (DCM) with a purity of 99.5% was purchased from Ling Feng Chemical Reagent Ltd (China) as the solvent to prepare β-carotene solution. Carbon dioxide (CO_2) with a purity of 99.95% was supplied by Rui Li Ltd (China).

2.2 Apparatus and Procedures

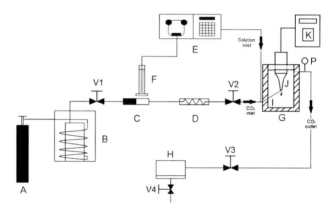

Figure 2   *Schematic diagram of the SEDS-EM apparatus*
A - CO_2 cylinder; B - chilling system (refrigerator); C - piston pump; D - heat exchanger; E - HPLC pump; F - carotene solution; G - high pressure vessel (with a heating tape and an insulation layer); H - back pressure regulator (BPR); I - coaxial nozzle; J - ultrasound horn (transducer); K - ultrasonic processor; P - pressure gauge; V1, V2, V3, V4-control valves.

The schematic diagram of SEDS-EM process for the production of resveratrol is shown in Figure 2. The experiments were carried out in the semi-continuous mode. The major parts of the apparatus include a supercritical CO_2 supply system, a solution feeding system and a high pressure precipitation vessel (G) which consists of an ultrasound generation system and a coaxial nozzle (I). CO_2 from the cylinder (A) was liquefied by a chilling system (B) and delivered constantly by a piston pump (C) through the heat exchanger (D) into the high pressure vessel (G, volume=120 cm^3). The pressure was maintained in the vessel by a back pressure regulator (H) and the temperature maintained by a heating tape with a thermocouple and an insulation layer. Once the pressure and temperature reached the desired values, the ultrasonic horn (J, made of Ti-6Al-4V, Φ_{tip}=15 mm) inside the vessel was turned on at the desired power supply (frequency=20 kHz), and the solution was injected inside the vessel through the capillary tube (inner part of the coaxial nozzle, Φ=50 μm). First, when the solution was contacted with the flowing CO_2 from the outer part of the coaxial nozzle, the solution jet mixed with CO_2 rapidly and was broken into small droplets immediately. Next, these small droplets were sprayed on the surface of the ultrasonic horn tip and were atomized into many smaller droplets. Particles precipitated swiftly from these droplets due to the removal of the solvent by supercritical CO_2. The diagram of the formation of droplets and nanoparticles in the SEDS-EM process is shown in Figure 3. After all the solution was injected, the ultrasonic processor was turned off and fresh CO_2 was continuously pumped into the vessel to flush the vessel to remove the residual organic solvent. The particles were collected by a filter placed at the CO_2 outlet of the vessel.

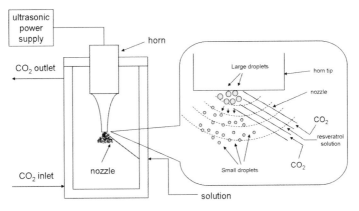

Figure 3   *The diagram of the formation of droplets in the SEDS-EM process*

2.3 Characterization

2.3.1 Morphology and size. Morphological characterization of samples was observed by a scanning electron microscopy (SEM, JEM-7401F, JEOL, Ltd., Japan). A small amount of specimen was placed on one surface of a double-faced adhesive tape that stuck to the sample support and coated with gold under vacuum condition for about 20 s to enhance the electrical conductivity of samples. The mean particle size and size distribution were measured with the Image-Pro software (version 6.0.0.260, Media Cybernetics, Inc.), using at least 500 particles for each experiment.

2.3.2 FT-IR analysis. Each sample of 1 mg was mixed with 100 mg KBr, and the mixtures were compacted to form disks. The infrared spectrum was recorded from 4400 to 500 cm^{-1} using a Perkin Elmer Fourier transform infrared spectrometer (Model: Spectrum 100).

3. RESULTS AND DISCUSSION

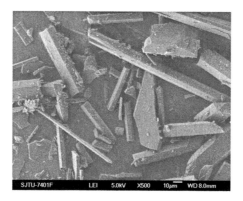

Figure 4   *Unprocessed resvertrol crystals*

Figure 5 *The SEM image of the resveratrol particles produced by SEDS technique*

The SEM image of the long-stick shape resveratrol crystals with the length more than 30 μm are shown in Figure 4. The morphology of the particles obtained by SEDS (Figure 5, T=34 °C, P=9.1 MPa, Cres=40 mg/ml, Flow rate of CO$_2$=3.0 kg/h, Flow rate of solution=1 ml/min) was changed to nanoflakes with the thickness of 20nm and the length or width of 4-8 μm. To evaluate the effect of the ultrasonic vibration in the antisolvent process, the SAS-EM processed resveratrol (T=34 °C, P=9.1 MPa, Cres=40 mg/ml, Flow rate of CO_2=3.0 kg/h, Flow rate of solution=1 ml/min, Power supply of ultrasound=240 W) was also conducted and the SEM image was shown in Figure 6. According to the particle size distribution (PSD) of this sample the particles are not uniform and the mean size of these particles is about 600 nm. In Figure 7, when SEDS-EM technique was applied in the production of resveratrol nanoparticles, the modification on the morphology and PSD of these particles are apparent. The particles appear more uniform with narrow PSD, and the mean particle size obviously decreased to 250 nm. The application of a coaxial nozzle and ultrasound provided enough suspension of the breakup of the solution jet to form extremely small droplets.[10] The combination of these two methods can improve the quality of the particles produced with the supercritical antisolvent process.

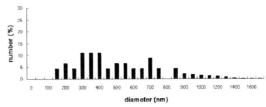

Figure 6   *The SEM image and PSD of the resveratrol particles produced by SAS-EM technique*

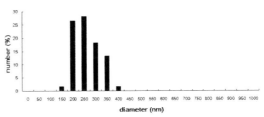

Figure 7   *The SEM image and PSD of the resveratrol particles produced by SEDS-EM technique*

However, in Figure 8, when the concentration of the resveratrol decreased from 40 mg/ml to 10 mg/ml (T=34 °C, P=9.1 MPa, Flow rate of CO_2 =3.0 kg/h, Flow rate of solution=1 ml/min, 120 W for SEDS-EM), the variation of the morphology and the particles size were not apparent between SEDS and SEDS-EM techniques. The supersaturation of the solution might be the most efficient factor to influence the particle precipitation instead of the droplet size.

In Figure 9, FT-IR results indicated that the structure of resveratrol did not change after the supercritical antisolvent process, even though the administration of the ultrasound in this process.

SEDS   SEDS-EM

Figure 8   *The SEM images of the resveratrol particles produced by SEDS and SEDS-EM*

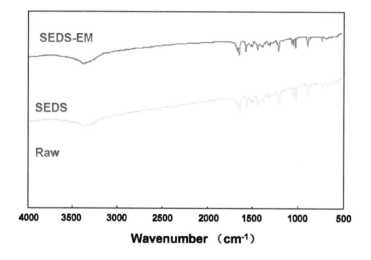

Figure 9   *The FT-IR spectroscopy of unprocessed and processed resveratrol*

4. CONCLUSIONS

A new SEDS-EM method was successfully developed to fabricate uniform and superfine resveratrol nanoparticles. Compared with SEDS and SAS-EM techniques, SEDS-EM can provide smaller and more uniform particles. This modified technique may promote the administration of supercritical antisolvent process in pharmaceutical industry.

Acknowledgement

This research is supported by Ministry of Science and Technology of the People's Republic of China, National Natural Science Foundation of China (2007AA10Z350, 20976103,).

References

1   B. Sjostrom, B. Kronberg and J. Carlfors. *J. Pharm. Sci.* 1993, **82**, 579.

2   E. Reverchon and A. Antonacci, *J Supercrit Fluids.* 2007, **39**, 444.

3   J. Bleich, P. Kleinebudde, and B. W. Mueller, *Int. J. Pharm.* 1994, **106**, 77.

4   M. Bahrami and S. Ranjbarian, *J. Supercrit. Fluids.* 2007, **40**, 263.

5   P. Chattopadhyay and R. B. Gupta, *Int. J. Pharm.* 2001, **228**, 19.

6   J. Baur JA and D.Sinclair, *Nat Rev Drug Discov*, 2006, **5** ,493.

7   M. Gentilli, J. Mazoit, and H. Bouazil, *Life Sciences*, 2001, **68**, 1317.

8   P. Kopp, *European Journal of Endocrinology*, 1998, **138**, 619.

9   H. C. Su, L. M. Hung, J. K. Chen, *AJP: Endocrinology and Metabolism*, 2006, **290**, 1339.

10 P. Chattopadhyay, R. B. Gupta, *AIChE J*, 2002, **48**, 235.

A SELF-ASSEMBLY MECHANISM OF CUPROUS OXIDE NANOPARTICLES IN AQUEOUS COLLOIDAL SOLUTIONS

Y. Bai, T. Yang, G. Cheng and R. Zheng

College of Nuclear Science and Technology, Beijing Normal University, Beijing, 100875 China

1 INTRODUCTION

Cuprous oxide and cupric oxide have been investigated for decades due to their unique semiconducting and optical properties. [1] As a P-type semiconducting material, the theoretical direct band gap of cuprous oxide is about 2.2 eV. [2] Cuprous oxide has a very long excited lifetime (about 10 μs), which can be used for photoluminescence. [3] Cuprous oxide has potential applications in solar cells, [4] nano-magnetic devices, [5] chemical industry, [6] sensors [7] and so on. It is also reported that cuprous oxide microspheres has been used as cathode material of lithium battery and photocatalyst in the visible light which led to photochemical decomposition of H_2O and generation of O_2 and H_2.[8] Structure-function relationship is the underlying motive for controlled fabrication of metal or semiconducting nano-materials. [9]

In the past few years, numerous Cu_2O nanostructures, including nanoplates, [10] nanocubes, [11] octahedra, [12] spherical particles, [13] nanoboxes, [14] and nanowires [15] were synthesized. The shape control and detailed crystal structure analysis of cuprous oxides have been performed on these Cu_2O nanocrystals. However, the growth mechanism, which is important for the controlled synthesis of Cu_2O nanocrystals, still needs a detailed investigation.

In this paper, we adopt an aqueous colloidal solution approach for the syntheses of monodispersed Cu_2O nanocubes, truncated nanocubes, cuboctahedra, nanosphere, and octahedra by adjusting experimental conditions, and explore the influence of experimental conditions on the morphology evolution of Cu_2O nanocrystals. Based on our observation and analysis, the growth mechanism of Cu_2O nanocrystals is elucidated.

2 EXPERIMENTAL

Cu_2O nanoparticles were prepared by the reaction of cupric acetate with ascorbic acid at different temperatures. Typically, 0.25 mmol (0.05 g) of cupric acetate and 0.45 mmol (in repeating unit) of polyvinylpyrrolidone (PVP) were dissolved in 100 ml deionized water, 0.75 mmol (0.132 g) of ascorbic acid was dissolved in 15 ml of deionized water and 0.005 mol (0.2 g) of sodium hydroxide was dissolved in 20 ml of deionized water to form homogeneous aqueous solutions. Then the as-prepared sodium hydroxide solution (0.25 mol/L) was added dropwise into the cupric acetate solution (2.5 mmol/L) under vigorous stirring at room temperature. The solution turned to be a blue suspension. Then the ascorbic acid aqueous solution (0.05 mol/L) was added into the above blue suspension at the speed of 3 drops per second under vigorous stirring. The color of the suspension changes from blue to green, finally, a reddish suspension was obtained after 10 minutes of reaction. The as-prepared products were centrifuged from the solution at 4000 rpm for 15 min using a Biofuges stratos centrifuger (Fisher Scientific). The derived products were washed with deionized water three times and dried in air for characterization. By changing the molar ratio of copper acetate to cupric acetate, the amount of surfactant, the reaction temperature and the stirring rate, Cu_2O nanoparticles with different amorphous were obtained.

Crystal structure of the Cu_2O nanoparticle was identified using a powder X-ray diffractometer (XRD) (PANalytical X' Pert), with Cu-Kα radiation (λ=1.5418Å) at 50 kV and 200 mA. The size and morphology of the Cu_2O nanoparticles was observed using a field emission scanning electron microscopy (SEM) (Hitachi S-4800, Japan) at 5 kV. Crystal structure and growth orientation of the Cu_2O nanoparticles was characterized with a high resolution transmission electron microscopy HRTEM (Philips FEI TECNAI F30) at 200 kV. The ultraviolet and visible light (UV-vis) absorption spectra was recorded using a Shimadzu UV-365 PC spectrophotometer.

3 RESULTS AND DISCUSSIONS

Figure 1 illustrates the microstructures of typical Cu_2O nanocubes synthesized. Figure 1a shows the SEM image of the samples. The nanoparticles exhibit typical cubic morphology with a mean edge length of 255 nm and a standard deviation of 35 nm. It could be observed that the corners of some nanocubes were truncated. Figure 1b shows the TEM image of a Cu_2O nanocube on the surface of a TEM grid. The inset shows the selected area electron diffraction (SAED) patterns obtained by directing the electron beam perpendicular to the square faces of the cube, The square symmetry of this pattern indicates that each Cu_2O nanocube was a single crystal bounded mainly by {100} facets. Figure 1c presents the XRD pattern of a reaction product in the 2θ range of 10-80°. The pattern could be distinctly indexed to a cubic phase with lattice constants a=4.267Å for Cu_2O (JCPDS No. 78-2076). It is worth noting that the ratio between the intensities of the (200) and (111) diffraction peaks was higher than the conventional value (1.28 versus 0.435), indicating that our nanocubes were abundant in {100} facets. Truncated corners of Cu_2O nanocubes correspond to {111} facets. UV-vis absorption spectra of cuprous oxide nanoparticles was demonstrated in Figure 1d. The characteristic absorption peak is due to plasma resonance excitation of copper atoms on the surface of nanocubes.[16] The absorption peak lies around 484 nm. The calculated direct band gap of our cuprous oxide is 2.56 eV, slightly larger than the theoretical direct band gap of 2.2 eV.

The morphology and the size of the nanoparticles could be tuned by adjusting the reaction conditions such as reaction temperature, reactant concentration, amount of surfactant, and stirring rates. In order to elucidate the effect of experimental parameter on the morphology of Cu_2O nanocrystals, we carry out four sets of experiments. In each set of experiments, only one parameter is varied while the other parameters are unchanged (i.e. the same parameters as those reported in the experimental section).

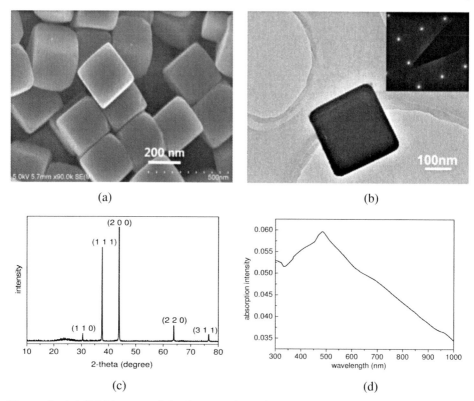

Figure 1   *(a) SEM images of Cu₂O nanocubes; (b) TEM images of Cu₂O nanocubes; (c) XRD pattern of as-prepared samples; (d) UV-visible spectrum recorded from solution of cuprous oxide nanocubes dispersed in deionized water at room temperature.*

Figure 2 shows various Cu_2O nanoparticles synthesized. In Figure 2, A1-A4 reveal the morphologies of Cu_2O nanoparticles synthesized with different amount of surfactant. Without surfactant, only irregular nanoparticles are obtained (A1). With the addition of surfactant, regular truncated nanocubes are produced. When the amount of surfactant increased from 0.18 mmol (A2) to 0.45 mmol (A3), the average particle size changed from 100 nm to 500 nm. The presence of PVP seems beneficial to the growth of Cu_2O nanocubes.

Temperature is a key factor in preparing uniform cubic Cu_2O particles. Exhibits B1-B4 in Figure 2 show the SEM images of Cu_2O nanocubes prepared at 30°C, 40°C, 50°C and 90°C, respectively. The morphology of particle changes from truncated cube to perfect cube with the temperature arising from 30°C to 50°C (B1-B3), particles in these three samples have smooth surface and are essentially mon-sized. Nanocubes prepared at 90°C are corroded seriously with uneven surfaces. With the rising of synthesis temperature, the average edge length of nanocubes exhibits a peak of 500 nm at 50°C.

The reactant concentration is another key factor that influences the morphology of Cu_2O nano-particles. At different copper acetate concentrations, most particles are truncated cube, except that the particle size varies form 100 nm to 1700 nm as the copper acetate concentration increases. However, with the addition of different amount of reducing agent, products with different morphologies are synthesized. Exhibits C1-C4 in Figure 2 are SEM images of Cu_2O nanocubes prepared with 0.25, 0.5, 0.75 and 1 mmol of ascorbic acid, respectively. When the amount of reducing agent was less than 0.25 mmol, the product is mainly comprised of $Cu(OH)_2$ nanowire. When the amount of reducing agent reaches 0.75 mmol, truncated cuprous oxide nanocubes with the average side length about 200 nm are obtained. The shape of cuprous oxide particle turns to rough octahedron with the amount of reducing agent increased to 1 mmol. When the amount of reducing agent was further increased to 1.25 mmol, there is no precipitation in solution after reaction.

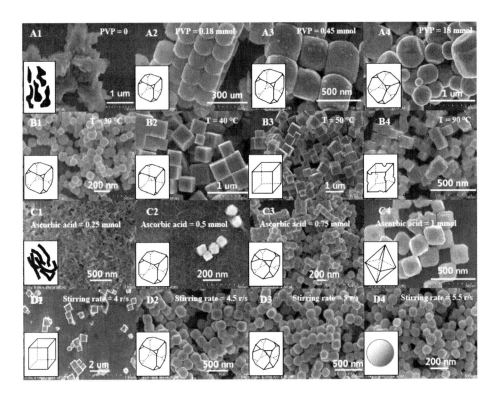

Figure 2 *SEM images of Cu_2O nanoparticles fabricated by different reaction conditions. Images in row A to D show the samples obtained at various reaction conditions: A- different amount of surfactant (0 to 0.018 mol); B – different temperatures (30 °C to 90 °C); C – different amounts of reducing agents (0. 246 mmol to 1 mmol); D- different stirring rates (4 r/s to 5.5 r/s).*

It is also interesting to find that the morphology of nanoparticles is very sensitive to the stirring rate used for synthesis. Small variation in the stirring rate will result in great changes in particle morphology. Exhibits D1-D4 in Figure 2 show the SEM image of Cu_2O nanoparticles synthesized at a stirring rate of 4 r/s, 4.5 r/s, 5 r/s and 5.5 r/s, respectively. The shape of nanoparticles changes from cube to truncated cube and then to spheres. Furthermore, the average particle size decreases from 750 nm to 150 nm with the increase of stirring rate.

4 DISCUSSION

It is speculated that Cu_2O tends to form nanocubes due to orientational crystallization mechanism. Cupric acetate could be dissolved in water and forms a uniform ionic solution. When NaOH is added in the solution, the Cu^{2+} react with OH^- and forms blue insoluble $Cu(OH)_2$ precipitate. While $Cu(OH)_2$ suspension react with ascorbic acid, $Cu(OH)_2$ will be reduced into reddish Cu_2O nanoparticles. The reactions can be expressed as:

$$Cu^{2+} + 2OH^- \rightarrow Cu(OH)_2 \downarrow \tag{1}$$

$$Cu(OH)_2 + C_6H_8O_6 \rightarrow Cu^{2+} + 2H_2O + C_6H_6O_6^{2-} \tag{2}$$

$$2Cu^{2+} + C_6H_6O_6^{2-} + H_2O \rightarrow Cu_2O + C_6H_6O_6 + 2H^+ \tag{3}$$

When Cu_2O precipitates from the solutions, they tend to aggregate and to reduce their total surface energy. When subsequent Cu_2O monomers precipitate from the solution, they tend to aggregate on the existing Cu_2O seeds and grow up. In the absence of a hard template, low-dimensional structures are governed by thermodynamic (e.g., temperature, reduction potential) and kinetic (e.g., reactant concentration, diffusion, solubility, reaction rate) parameters.

It is reported that PVP has the selective adsorption properties to specific crystal planes, and could be used to kinetically control the growth of single crystalline nanocubes. It is also found that surfactant plays an important role on morphology and size of nanocrystals. Without PVP surfactant, the structure of products is polycrystalline cluster without certain morphology, which indicates Cu_2O nanoparticles aggregate in a disorder manner. When PVP was added into solution, monodispersed colloidal solutions of Cu_2O nanocrystals with regular polyhedral shapes and bound entirely by {100} and {111} facets of the fcc crystal lattice are obtained. It is observed that the Cu_2O nanocubes with larger particle sizes are obtained at higher PVP concentrations, which indicates that the existence of PVP promotes the growth of Cu_2O nanocubes. It is anticipated that the interaction between PVP and Cu_2O monomer raises the critical nucleation energy, which suppresses the nucleation rate and makes Cu_2O nanocubes grow faster.

As an important thermodynamic parameter, temperature exhibits a significant influence on the morphology of nanocrystals. At the room temperature, with the aid of surfactants (it is believed that the selective interaction between PVP and {100} planes of Cu_2O could greatly reduce the growth rate along the <100> direction), [17] truncated Cu_2O nanocubes are synthesized. From energy point of view, this structure is stable. Because the corners of nanocube have larger specific surface area, the corners of nanocubes are corroded and truncated along the {111} planes. As the synthesis temperature increases, diffusion and migration rate of the ions and monomers increase, the nanocubes grow fast along the <111> directions. Hence, the morphology of product changes gradually from truncated nanocube to nanocube. At the same time, particle size increased significantly. However, further increase in synthesis temperature will result in heavy corrosion of cube surfaces.

Ascorbic acid is a mild reducing agent, it could reduce $Cu(OH)_2$ and forms Cu_2O nanocubes. At the same time, Ascorbic acid works as a vinylogous carboxylic acid. Excessive ascorbic acid will corrode the Cu_2O nanocubes via a dismutation reaction. Since the {100} planes of nanocubes are selective adsorbed by the capping agent PVP, {111} planes become candidates for selective corrosion. With the increase of the amount of ascorbic acid, the morphology of Cu_2O nanocubes varies from truncated cube to cuboctahedra and further becomes octahedron. Further increasing the amount of ascorbic acid will cause the completely corrosion of Cu_2O nanoparticles, and the suspension turns to be transparent solution again. This finding is of great value, because previous report indicates that {111} planes of Cu_2O particles have high catalytic activity.[18 19] In addition to the thermodynamic and kinetic parameters, the shape of Cu_2O nanocrystal is sensitive to the stirring rate. In fact, the rapid stirring rate actually hinders the agglomeration of particles and prevents the growth of crystals.

Moreover, the high stirring rate will enhance the probability and energy of collision between particles and wall. Mechanical wear and erosion make corners and edges of Cu_2O nanocubes disappear gradually. When the stirring rate is 4 r/s, perfect cuprous oxide nanocubes with average edge length of 750 nm are synthesized. When the stirring rate is increased to 5 r/s, truncated cubes with average diameter of 250 nm are formed in solution. When the stirring rate reaches 6 r/s, only nanospheres of 150 nm diameter are produced.

5 CONCLUSION

In this paper, cuprous oxide nanoparticles with different morphologies have been successfully synthesized using a wet-chemical approach in aqueous solution. XRD and TEM results indicate that the cuprous oxide particles are truncated nanocubes bounded by {100} and {111} facets. At various synthesis conditions, Cu_2O nanocubes, truncated nanocubes, cuboctahedra, nanosphere, and octahedra are synthesized. The growth mechanism of these nanoparticles is interpreted as the combined effects of selective growth, selective corrosion and mechanical erosion of Cu_2O nanocrystals. This implies that Cu_2O nanoparticles with well-controlled shapes, sizes, and structures can be obtained with optimized thermodynamic and kinetic parameters.

References

1   I. Grozdanov, *Mater. Lett.*, 1994, **19**, 281.
2   P.B. Ahirraoa, B.R. Sankapalb and R.S. Patil, *J. Alloy. Compd.*, 2011, **59**, 5551.
3   R.M. Habiger and A. Compaan, *Solid State Commun.*, 1976, **18**, 1531.
4   L. F. Gou and C. J. Murphy, *Nano Lett.*, 2003, **3**, 231.
5   A. Ahmed, N. S. Gajbhiye and S. Kurian, *J. Solid State Chem.*, 2010, **183**, 2248.
6   W. Z. Wang and W. J. Zhu, *J. Phys. Chem. B*, 2006, **110**, 13829.
7   J. Liu, S. Wang, Q. Wang and B. Geng, *Sensor Actuat. B*, 2009, **143**, 253.
8   M. Hara, T. Kondo, M. Komoda, S. Ikeda, K. Shinohara, A. Tanaka, J. N. Kondo and K. Domen, *Chem. Commun.*, 1998, **3**, 357.
9   Y. Sun and Y. Xia, *Science*, 2002, **298**, 2176.
10 C.-H. Kuo and M. H. Huang, *Nano Today*, 2010, **5**, 106.
11 Z. Wang, H. Wang, L. Wang and L. Pan, *J. Phys. Chem. Sol.*, 2009, **70**, 719.
12 X. Zhang, Y. Xie, F. Xu, X. Liu and D. Xu, *Inorg. Chem. Commun.*, 2003, **6**, 1390.
13 H. Xu, W. Wang and W. Zhu, *Micropor. Mesopor. Mat.*, 2006, **95**, 321.
14 L. Huang, F. Peng, H. Yu and H. Wang, *Mater. Res. Bull.*, 2008, **43**, 3407.
15 H. S. Shin, J. Y. Song and J. Yu, *Mater. Lett.*, 2009, **63**, 397.
16 M. Y. Shen, T. Yokouchi, S. Koyama, and T. Goto, *Phys. Rev. B*, 1997, **56**, 13066.
17 E. Ko, J. Choi, K. Okamoto, Y. Tak and J. Lee, *Chem. Phys. Chem.*, 2006, **7**, 1505.

18 P. He, X. Shen and H. J. Gao, *Colloid Interface Sci.*, 2005, **284**, 510.
19 H. Xu, W. Wang and W. Zhu, *J. Phys. Chem. B,* 2006, **110**, 13829.

ONE-STEP SYNTHESIS OF MESOPOROUS SULFATED ZIRCONIA NANOPARTICLES WITH ANION TEMPLATE

L. Y. Zhang,[1] H. Wang,[2] S. W. Xu,[1] C. Y. Han,[1] Y. Y. Zhang,[1] D. Q. Du,[1] J. Y. Li[1] and Y. M. Luo[1,2]

[1] Faculty of Environmental Science and Engineering, Kunming University of Science and Technology, Kunming 650093, China. E-mail: luoyongming@tsinghua.org.cn.
[2] Faculty of Metallurgical and Energy Engineering, Kunming University of Science and Technology, Kunming 650093, China.

1 INTRODUCTION

Sulfated zirconia has attracted a lot of attention because of its high catalytic activity for the conversion in many important industrial and organic reactions.[1-3] Mesoporous sulfated zirconia (MSZ) exhibits much higher catalytic activity than conventional sulfated zirconia (CSZ) due to its high surface area, uniform pore size and high porosity.[4-7] Therefore, many researches have been devoted to synthesize MSZ with high surface area by using various methods.[8-25] In general, these synthesis routes of sulfated zirconia can be classified into one-step and two-step, while most of them were concentrated on two-step synthesis. Compared with two-step synthesis, one-step synthesis route has attracted much attention due to the advantages of avoiding impregnation (sulfating) step and simplifying the overall synthesis procedure to a great extent.[1,21] Schüth et al[23] reported that MSZ has been synthesized using cation template CTAB as a structure-directing agent via a one-step method for the first time. Cheng et al. also reported MSZ has been synthesized using CTAB and $Zr(SO_4)_2$.[12] Fripiat et al. reported mesoporous zirconia rather than MSZ has been synthesized using SDS as a template for the first time.[26] Ozawa et al. reported an one-step route for the synthesis of sulfated zirconia powder using $Zr(OC_3H_7)_4$ and SDS as a zirconium source and a template, respectively. However, mesostructure, confirmed by their small angle XRD, was not formed at all. More importantly, BET surface area of the sample after calcination is only 24 m^2/g.[27]

Herein, we report a new facile one-step route for the synthesis of mesoporous sulfated zirconia nanoparticles using SDS both as a template and a sulfating agent for the first time, and the formation mechanism is investigated and discussed in detail.

2 EXPERIMENTAL

$Zr(OC_3H_7)_4$ and Sodium dodecyl sulfate ($C_{12}H_{25}OSO_3Na$, SDS) were used as a zirconium source and as a structure-directing agent, respectively. In a typical synthesis batch, 13.5g of SDS was added into 200 ml of deionized water with vigorously stirring at room temperature (RT). After SDS was dissolved completely, calculated $Zr(OC_3H_7)_4$ was added into the mixture solution to keep the molar ratio of Zr/S between 0.25 and 4.0. Thereafter, the pH value of mixture solution was adjusted by using HCl (1.0 M) to 3.0 and the mixture solution was stirred for 0.5 h. Then, the mixture was aged at RT for 20 h and thermally treated at 80 °C for 5 days under static conditions in turn. Subsequently, the reaction product was filtered, washed with the mixture solution of deionized water and isopropanol and dried at 110 °C for 12-24 h. Finally, the sample was calcined at various temperatures (400 °C, 550 °C, 650 °C and 750 °C) in air for 3 h. The sample synthesized by this route was designated as MSZNP$_{(x)}$, where "x" is calcination temperature, and the as-synthesized sample without calcination was nominated as MSZNP$_{(AS)}$.

For comparison purpose, sulfated zirconia was prepared by using traditional two step method. The synthesis procedures of Zr(OH)$_4$ are similar to those of mesoporous sulfated zirconia nanoparticles except that SDS was not added into synthesis media. The dried Zr(OH)$_4$ was ground and was added into H$_2$SO$_4$ solution (1.0 M) with stirring for 2 h. Then, the sulfated product was filtered and dried at 110 °C for 12-24 h. Finally, the sample was calcined at 550 °C in air for 3 h. The sulfated zirconia sample synthesized by this route was designated as CSZ$_{(550)}$.

3 RESULTS AND DISCUSSION

Table 1 *BET surface area of MSZNP$_{(AS)}$, MSZNP$_{(400)}$, MSZNP$_{(550)}$, MSZNP$_{(650)}$ and CSZ$_{(550)}$*

Sample	MSZNP$_{(AS)}$	MSZNP$_{(400)}$	MSZNP$_{(550)}$	MSZNP$_{(650)}$	CSZ$_{(550)}$
Surface area/ m^{2}g^{-1}	336	204	162	108	97

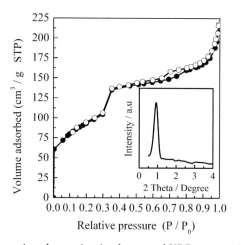

Figure 1 *N$_2$ adsorption-desorption isotherm and XRD pattern (inset) of MSZNP*

N$_2$ adsorption-desorption isotherm of MSZNP$_{(AS)}$ is found to be of type IV. Three well distinguished regions of the adsorption isotherm are evident: (i) monolayer-multilayer adsorption (P/P$_0$ < 0.1), (ii) capillary condensation (P/P$_0$= 0.3-0.4), and (iii) multilayer adsorption on the outer particle surfaces (P/P$_0$ >0.9). The presence of a single diffraction peak in 2θ region blow 2° for MSZNP$_{(AS)}$ is indicative of a disordered mesostructure.[28,29] Similar N$_2$ adsorption-desorption isotherms are observed for MSZNP$_{(400)}$ MSZNP$_{(550)}$ and MSZNP$_{(650)}$, while the step sharpness at P/P$_0$ = 0.3-0.4 decreases with the calcination temperature (pattern not shown), indicating the presence of mesostructure after calcination. BET surface areas of MSZNP$_{(AS)}$, MSZNP$_{(400)}$, MSZNP$_{(550)}$, MSZNP$_{(650)}$ and CSZ$_{(550)}$ are summarized in Table 1. It is clear that BET surface areas of MSZNP$_{(550)}$ is much higher than that of CSZ$_{(550)}$ and sulfated zirconia prepared with other conventional two-step method.[16,21]

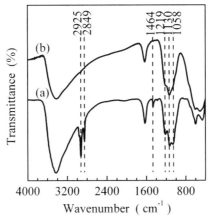

Figure 2 *FT-IR patterns of (a) MSZNP$_{(AS)}$ and (b) MSZNP$_{(550)}$*

Three bands centered at 1058, 1130 and 1219 cm^{-1} attributed to characteristic IR vibrations of chelating bidentate sulphate ion, are observed for MSZNP$_{(AS)}$ and MSZNP$_{(550)}$.[22,30] The bands centered at 1464, 2849 and 2925 cm^{-1} are well-expressed in the spectrum of MSZNP$_{(AS)}$, but not for MSZNP$_{(550)}$. The former can be assigned to the asymmetric stretching vibrations (δ_{as}) of methyl groups (CH$_3$),[31,32] and the latter two are ascribed to the characteristic vibration of asymmetric and symmetric vibrations (δ_{as} and δ_s) for methylene groups (CH$_2$).[32,33] The difference is likely because SDS was completely removed, while the SO$_4^{2-}$ species were directly incorporated into zirconia.

Figure 3 *TEM image and EDS profile (inset) of MSZNP$_{(550)}$.*

From the TEM image, nanoparticles with average diameter of 7.2 nm can be easily identified. It is clear that three elements (Zr, O and S) are detected in the EDS profile of MSZNP$_{(550)}$, indicating these nanoparticles are sulfated zirconia rather than zirconia. Further quantitative analysis shows that sulfur content of MSZNP$_{(550)}$ is 3.74%, which is comparable to that of CSZ$_{(550)}$ prepared with conventional method.

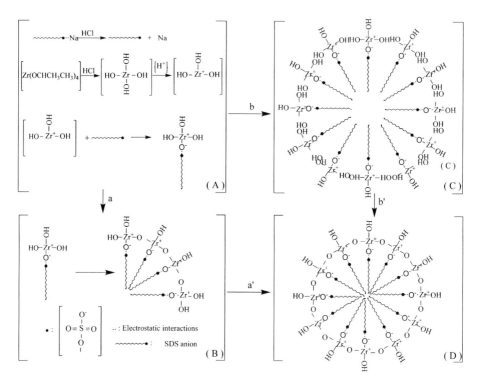

Figure 4   *Schematic diagram of MSZNP formation.*

The mechanism of the formation MSZNP is illustrated in Figure 4. The template SDS was dissolved in water to yield $C_{12}H_{25}OSO_3^-$ anion, which contains one hydrophobic and one hydrophilic head. It is well-documented that the isoelectric point of ZrO_2 is about 5.6 (pH$_{ZPC}$ ≈ 5.6).[34-36] This means that zirconium species are charged with positive ions when pH is lower than 5.6. Therefore, when the pH was adjusted to 3.0 by adding HCl into synthesis media drop-wise, positive zirconium species were formed via hydrolysis of $Zr(OC_3H_7)_4$. Then, the positive zirconium species interacted with the hydrophilic head of $C_{12}H_{25}OSO_3^-$ via electrostatic interaction to form inorganic and organic hybrids (Figure 4A). After that, the composites contained template and zirconium species (Figure 4D) were obtained via two possible pathways. In the first pathway (step a and a'), the self-assembly step (a') to form circular structure occurs only after sufficient propagation of the polycondensation hybrids (Figure 4B). In the second pathway (step b and b'), the micellar arrays structure (Figure 4C) was formed prior to subsequent condensation (b'). The template SDS was removed when the resulting product was calcined, as evidenced by FT-

IR (Figure 2). At the same time, the SO_4^{2-}, originated from the hydrophilic moieties of SDS, was successfully incorporated into zirconia due to the strong interaction between the zirconium species and sulfuric species.

4 CONCLUSIONS

Mesoporous sulfated zirconia nanoparticles have been synthesized by employing sodium dodecyl sulfate (SDS) both as a template and a sulfating agent via one-step route. It was well documented that the SO_4^{2-} anion, originated the hydrophilic head of template SDS, can be directly incorporated into zirconia to form super-acid, and the corresponding formation mechanism was illustrated.

Acknowledgements

This research work was supported by National Natural Science Foundation of China (Grant No. 20867003 and 51068010), Postdoctoral Science Foundation Funded Project (Grant No. 20100471686), Young Academic and Technical Leader Raising Foundation of Yunnan Province (Grant No. 2008py010).

References

1   B. M. Reddy and M. K. Patil, *Chem. Rev.*, 2009, **109**, 2185.
2   G. D. Yadav and J. J. Nair, *Micropor. Mesopor. Mater.*, 1999, **33**, 1.
3   X. M. Song and A. Sayari, *Catal. Rev. Sci. Eng.*, 1996, **38**, 329.
4   M. A. Risch and E. E. Wolf, *Appl. Catal. A: Gen.*, 1998, **172**, L1.
5   M. Risch and E. E. Wolf, *Catal. Today*, 2000, **62**, 255.
6   E. Zhao, S. E. Hardcastle, G. Pacheco, A. Garcia, A. L. Blumenfeld and J. J. Fripiat, *Micropor. Mesopor. Mater.*, 1999, **31**, 9.
7   X. Yang, F. C. Jentoft, R. E. Jentoft, F. Girgsdies and T. Ressler, *Catal. Lett.*, 2002, **81**, 25.
8   M. K. Mishra, B. Tyagi and R. V. Jasra, *Ind. Eng. Chem. Res.*, 2003, **42**, 5727.
9   M. Risch and E. E. Wolf, *Stud. Surf. Sci. Catal.*, 2000, **130**, 2381.
10 M. Risch and E. E. Wolf, *Appl. Catal. A: Gen.*, 2001, **206**, 283.
11 Y.-W. Suh and H.-K. Rhee, *Stud. Surf. Sci. Catal.*, 2002, **141**, 289.
12 Y.-W. Suh, J-W. Lee and H.-K. Rhee, *Catal. Lett.*, 2003, **90**, 103.
13 Y.-W. Suh, J.-W. Lee and H.-K. Rhee, *Appl. Catal. A: Gen.*, 2004, **274**, 159.
14 V. G. Devulapelli and H.-S. Weng, *Catal. Commun.*, 2009, **10**, 1711.
15 M. Lutecki, O. Solcova, S. Werner and C. Breitkopf, *J. Sol-Gel. Sci. Technol.*, 2010, **53**, 13.
16 D. J. McIntosh and R. A. Kydd, *Micropor. Mesopor. Mater.*, 2000, **37**, 281.
17 M. Signoretto, A. Breda, F. Somma, F. Pinna and G. Cruciani, *Micropor. Mesopor. Mater.*, 2006, **91**, 23.
18 Y.-Y. Huang, T. J. McCarthy and W. M. H. Sachtler, *Appl. Catal. A: Gen.*, 1996, **148**, 135.
19 Y. Y. Sun, L.Yuan, W. Wang, C.-L. Chen and F.-S. Xiao, *Catal. Lett.*, 2003, **87**, 57.
20 S. K. Das, M. K. Bhunia, A. K. Sinha and A. Bhaumik, *J. Phys. Chem. C*, 2009, **113**, 8918.

21 S. Melada, S. A. Ardizzone and C. L. Bianchi, *Micropor. Mesopor. Mater.*, 2004, **73**, 203.

22 Y. Y. Sun, S. Q. Ma, Y. C. Du, L. Yuan, S. C. Wang, J. Yang, F. Deng and F.-S. Xiao, *J. Phys. Chem. B.*, 2005, **109**, 2567.

23 U. Ciesla, S. Schacht, G. D. Stucky and F. Schüth, *Angew. Chem. Int. Ed.*, 1996, **35**, 541.

24 U. Ciesla, M. Fröba, G. D. Stucky and F. Schüth, *Chem. Mater.*, 1999, **11**, 227.

25 S.-Y. Chen, L.-Y. Jang and S. F. Cheng, *J. Phys. Chem. B.*, 2006, **110**, 11761.

26 G. Pacheco, E. Zhao, A. Garcia, A. Sklyarov and J. J. Fripiat, *Chem. Commun.*, 1997, 491.

27 M. Ozawa, D. Yokoi and S. Suzuki, *J. Mater. Sci. Lett.*, 2003, **22**, 1543.

28 P. T. Tanev and T. J. Pinnavaia, *Science*, 1995, **267**, 865.

29 Y. M. Luo, Z. Y. Hou, D. F. Jin, J. Gao and X. M. Zheng, *Mater. Lett.*, 2006, **60**, 393.

30 F. Babou, G. Coudurier and J. C. Vedrine, *J. Catal.*, 1995, **152**, 341.

31 V. Lochař, *Appl. Catal. A: Gen.*, 2006, **309**, 33.

32 H. G. Bernal, L. C. Caero, E. Finocchio and G. Busca, *Appl. Catal. A: Gen.*, 2009, **369**, 27.

33 V. Sánchez Escribano, C. del Hoyo Martínez, E. Fernández López, J. M. Gallardo Amores and G. Busca, *Catal. Commun.*, 2009, **10**, 861.

34 K. D. Hristovski, P. K. Westerhoff, J. C. Crittenden and L. W. Olson, *Environ. Sci. Technol.*, 2008, **42**, 3786.

35 K. Hristovski, A. Baumgardner and P. Westerhoff, *J. Hazard. Mater.*, 2007, **147**, 265

36 R. Greenwood and K. Kendall, *J. Eur. Ceram. Soc.*, 1999, **19**, 479.

RAPID AND CONTINUOUS SYNTHESIS OF LiFePO$_4$ AND LiFePO$_4$/C NANOPARTICLES IN SUPERCRITICAL WATER

X. Song and Y.P. Zhao

School of Chemistry and Chemical Engineering, Shanghai Jiaotong University, 800 Dongchuan RD. Shanghai 200240, China
Email: ypzhao@sjtu.edu.cn

1 INTRODUCTION

LiFePO$_4$ can be used as a cathode material for rechargeable lithium-ion batteries.[1] However, its application is limited due to the relatively poor ionic and electronic conductivity.[2] It has been reported that the low conductivity of LiFePO$_4$ can be improved by synthesizing superfine particles, carbon coating and metal-ion doping.[3-5] The traditional production of LiFePO$_4$ is usually time-consuming, energy-inefficient, and costly.

Supercritical water (SCW) has recently gained great interest as a green and controllable media for synthesis of LiFePO$_4$.[6-8] In this work，a rapid and continuous synthesis of LiFePO$_4$ nanoparticles and LiFePO$_4$/C in supercritical water was explored. The effects of temperatures, residence time and the organic substances on the formation of LiFePO$_4$ particles were investigated. Carbon coating of LiFePO$_4$ was also examined.

2 MATERIALS AND METHOD

2.1 Reagents

Water-soluble starch, ferrous sulfate (FeSO$_4$·7H$_2$O), lithium hydroxide (LiOH·H$_2$O), phosphoric acid (o-H$_3$PO$_4$), Citric acid (C$_6$H$_8$O$_7$), Glucose (C$_6$H$_{12}$O$_6$), and vitamin C were analytical grade and purchased from Shanghai Chemical Co., Shanghai, China.

2.2 Apparatus and Experimental Procedure

The synthesis of LiFePO$_4$ and LiFePO$_4$/C was carried out in a continuous tubular reactor as shown schematically in Figure 1. Three kinds of the reaction solutions were delivered continuously into the tubular reactor by three pumps, respectively. First, the solution of FeSO$_4$ and o-H$_3$PO$_4$ containing a known amount of vitamin C was premixed with the solution of LiOH using a pre-mixer. Secondly, the mixture solution was mixed with the preheated deionized water and was heated rapidly. Thirdly, the hot solution was further heated to the desire temperature through the reactor. The pressure of the system was adjusted by the back pressure valve. The resultant solution coming out from the reactor was quenched by the cooling water and collected in the vessel via the back pressure valve. Finally, LiFePO$_4$ was separated from the produced solution and dried at 100°C for 2 h for further characterization. During reaction, the flowrate ratio of the mixture solution of FeSO$_4$ and H$_3$PO$_4$, the solution of LiOH and the deionized water was fixed at 1:1:5. The temperature of both the preheater and the reactor was measured using thermocouples.

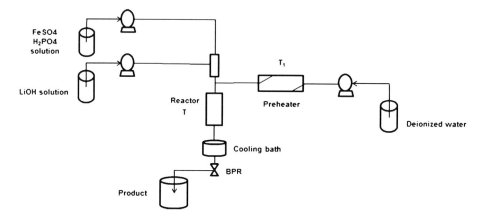

Figure1 *Schematic diagram of a continuous tubular reactor*

2.3 Characterization

The dried particles were analyzed using SEM, TEM, DLS and XRD. Transmission Electron Microscope (TEM, A JEOL JEM2010F) was used to obtain electron diffraction patterns of the samples. Particle sizes were estimated from SEM micrographs (JOEL, JEM-7401F) and DLS（Zetasizer Nano S）. The X-ray diffraction patterns of the samples were obtained with XRD (Model PW 1800, Phillips) using a Cu K radiation source.

2.4 Experimental Design

Nine experiments were performed with glucose and citric acid added to one of them, separately. The experimental conditions are summarized in Table 1.

Table 1 *Summary of synthesis conditions for LiFePO$_4$*

Sample	Temperature at preheater (℃)	Temperature at reactor (℃)	residence time (s)	Feed concentration (M)			Pressure (MPa)
				Fe^{2+}	PO$_4$$^{3-}$	Li$^+$	
L1	355	380	80	0.015	0.015	0.045	27
L2	355	380	60	0.015	0.015	0.045	27
L3	355	380	40	0.015	0.015	0.045	27
L4	405	380	80	0.015	0.015	0.045	27
L5	405	380	60	0.015	0.015	0.045	27
L6	405	380	40	0.015	0.015	0.045	27
L7	405	380	60	0.03	0.03	0.09	23
L8[a]	405	380	60	0.03	0.03	0.09	23
L9[b]	405	380	60	0.03	0.03	0.09	23

[a]L8: addition of Glucose during the SCW synthesis.
[b]L9: addition of Citric acid during the SCW synthesis.

3 RESULTS AND DISCUSSION

3.1 Effects of Temperature and Residence Time on the Formation of LiFePO₄

Figure 2 shows the XRD spectra of the samples listed in Table 1. It can be found that when the temperature of the preheater is 405 ℃, all diffraction peaks can be attributed to the orthorhombic olivine type phase $LiFePO_4$ (L4-L8) though the residence time is only 40 s (L6), which confirms that the samples obtained were pure $LiFePO_4$. However, when the temperature of the preheater is 355 ℃, the crystallinity of the samples is low (L1, L2 and L3) even though the residence time is 80 s (L3). Figure 3 shows that the particle size of $LiFePO_4$ is smaller in 40 s than in 80 s when the temperature of the preheater is 405 ℃. This suggests that a higher temperature and a shorter residence time promote the formation of the smaller particle size with higher crystallinity.

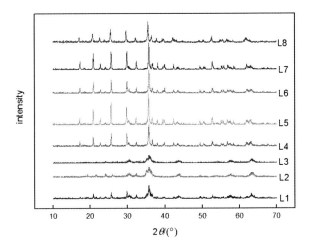

Figure 2 *X-ray diffraction patterns of LiFePO₄ prepared under the conditions listed in Table 1*

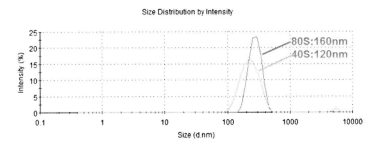

Figure 3 *DLS patterns of LiFePO₄ at different residence time (the temperature is 405 ℃)*

3.2 Effects of Temperature and Residence Time on the Morphology of LiFePO₄

Typical SEM images of the samples L1 to L6 are shown in Figure 4. It can be seen that the morphology of the LiFePO₄ particles synthesized at 355℃ is different from that at 405℃. In addition, irregular shape of particles can be found at 355 ℃. LiFePO₄ particles synthesized at 405 °C (L3, L4 and L5) are almost spherical at different residence time. However, the particle sizes are smaller and more uniform when the residence time is shortened, which is consistent with the DLS analysis shown in Figure 3. It is believed that the particles could be aggregated to form irregular clusters when the residence time is longer.

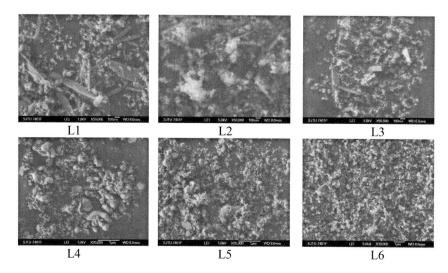

Figure 4 *Representative SEM micrographs of the samples (L1- L6)*

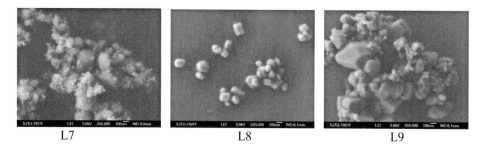

Figure 5 *Representative SEM micrographs of LiFePO₄*

3.3 Effects of Organic Substances on the Size and Morphology of LiFePO₄

In order to investigate whether the organic substances affect the morphology of LiFePO₄, glucose and citric acid were chosen as two kinds of model substances to be added into the reaction solution. Figure 5 shows the SEM images of the samples L7, L8 and L9. The particles are almost spherical. Comparing Fig.5b and 5c with Fig.5a, it can be found that glucose and citric acid obviously affect the morphology of the LiFePO$_4$ particles. Furthermore, smaller size and more uniform size distribution of particles were obtained when the glucose was used, compared to those obtained with citric acid.

3.4 Carbon Coating of LiFePO$_4$

In order to explore carbon coating of LiFePO$_4$, a known amount of the samples of L7 was dispersed into the solution of glucose for some time firstly and then filtered out and dried at 100 ℃, and finally calcined for two hours at the temperature of 600 ℃ under N$_2$ atmosphere. The SEM and TEM images of the calcined sample are shown in Figure 6. It can be seen clearly that the particles of LiFePO$_4$/C were spherical and it was coated by a carbon layer which is about 2-3 nm thick.

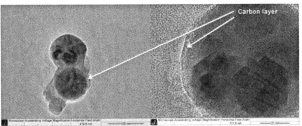

Figure 6 *Representative SEM and TEM images of carbon coated LiFePO4*

4 CONCLUSIONS

The continuous hydrothermal technique was successfully applied to synthesize LiFePO$_4$ and LiFePO$_4$/C nanoparticles. It was found that the temperature of the preheater had a significant effect on particle crystallinity and morphology. Shorter residence time promoted the formation of smaller spherical particles at high temperatures. Organic substance can affect crystal growth and morphology of particles. The smaller size and more uniform size distribution of LiFePO$_4$ nanoparticles were synthesized when glucose was used. Carbon can be coated on the surface of the LiFePO$_4$ successfully with a layer of 2-3 nm thick.

References

1. A.K. Padhi, K.S. Nanjundaswamy and J.B. Goodenough, *J. Electrochem.Soc.*, 1997,**144**: 1188.
2. L. N. Wang and K.L. Zhang, *J. Power Sources*. 2007, **167**, 200.
3. M.M. Doeff, J.D. Wilcox, R. Kostecki and G.J. Lau, *Power Sources*. 2006, **163,** 180.
4. Y. Xia, M. Yoshio and H. Noguchi, *Electrochim. Acta*. 2006, **52**, 240.
5. G. Liang, L. Wang, X. Ou, X. Zhao and S. Xu, *J. Power Sources*, 2008,**184**,538.
6. J. Lee and A.S. Teja, *Mater. Lett.*, 2006, **60**, 2105.
7. C.B. Xu, J. Lee and A.S. Teja, *J. Supercrit. Fluids*, 2008, **44**, 92.
8. W.L. Yu, Y.P. Zhao and Q.L. Rao, *Chinese Journal of Chemical Engineering*, 2009,**17**,174.

HYDROTHERMAL SYNTHESIS OF TIO$_2$ NANOPARTICLES: PROCESS MODELLING AND EXPERIMENTAL VALIDATION

M. Chen, C. Y. Ma, T. Mahmud, T. Lin and X. Z. Wang

Institute of Particle Science and Engineering, School of Process, Environmental and Materials Engineering, University of Leeds, Leeds LS2 9JT, U.K.

1   INTRODUCTION

Continuous hydrothermal synthesis (CHS) method has been attracting significant attention in recent years as a simple, efficient and environmental friendly process for nano-material synthesis.[1-3] The process uses supercritical water as a reagent that reacts with metal salt solution to produce nano-particles in a reactor. Particle size is a vital property in CHS process, because it is closely related to product quality and functionality. Furthermore, controlling particle size can help to enhance solution mixing and improve precipitation efficiency.[4] An important step in size control is to use mathematical models to simulate the particle size distribution (PSD) at different process operating conditions. Only a few works can be found in literature in this area.[5-8] In our previous work, a PB model was developed to predict PSDs of nano-particles synthesised by CHS process in a mixed suspension mixed product removal (MSMPR) reactor.[9]

In this paper, A CHS experiment rig was built to produce nano-size TiO$_2$ particles. The manufactured nano-particles were characterized using scanning electron microscopy (SEM) and image analysis techniques. A population balance model with revised parameters was developed and validated using the experimental data.

2   POPULATION BALANCE MODEL

Population balance (PB) modelling is an effective simulation tool for particulate processes. It estimates the dynamic evolution of PSD as a function of process operating conditions.[10] In our previous work, a PB model was developed for estimating nano-particle PSD produced in CHS process.[9] In this section, a modified PB model for CHS synthesis in a high pressure continuous stirred tank reactor (CSTR) is introduced.

For the CSTR, it is assumed that the system is in homogenous condition and only particle diameter is considered as the model internal coordinate. The PB equation is given as:

$$\frac{dn(x,t)}{dt} + n(x,t)\frac{dG(x)}{dx} = Agg(x) + B(x) \tag{1}$$

$$Agg(x) = \frac{1}{2}\int_0^x (x-x',t)n(x',t)q(x-x',x')dx - n(x,t)x\int_0^x n(x',t)q(x,x')dx' \tag{2}$$

where, t is the time (s); x is the particle diameter (m); $n(x,t)$ is the number density of particles with size range from x to $(x+dx)$ at time t; $G(x)$ is the growth rate of particles (m/s), $B(x)$ is the nucleation rate of particles (number of nuclei formed/m^3s), Agg(x) is the aggregation source term and $q(x,x')$ is the aggregation kernel.

In order to obtain the species concentrations for calculating supersaturation, nucleation and growth rates, the following mass balance equation is used:

$$\frac{dc}{dt} = \frac{W}{V}(kc_f - \bar{c}) \tag{3}$$

where, W is the flow-rate (ml/min), V is the reactor volume (ml), c_f is the feed precursor concentration (wt%), c is the solute concentration in liquid phase (wt%), ρ is the density of water (kg/m^3), k is the reaction rate constant of hydrothermal reaction and $\bar{c}$ is the combined of concentration of aqueous and solid phase product molecules (wt%), which is given as:

$$\bar{c} = c + \frac{\rho_s}{\rho} \times \frac{4}{3}\pi\int_0^\infty x^3 n(x,t)dx \tag{4}$$

Detailed equations for modelling k, $G(x)$, $B(x)$ and $q(x,x')$ can be found in pervious work.[9] A commercial modelling programme gPROMS (general process modelling system, PSE Ltd.) was used for model development and simulation. The PB equation was solved using the backward finite difference discretized method.

3 EXPERIMENTAL

3.1 Materials

50wt% TiBALD (Titanium (Ⅳ) bis(ammonium lactate) duhydroxide) from Aldrich Chemical Co. were used without further purification. Reference TiO_2 powders (Aeroxide® P25) were purchased from Degussa Co.

3.2 Apparatus and Procedure

The experimental apparatus used for the CHS process is shown in Figure 1. Two high-pressure liquid chromatography pumps were used to supply the metal salt precursor solution and distilled water to the reactor. The distilled water was heated-up and transformed to supercritical water (ScH_2O) inside a Separex supercritical flow oven system. The system pressure was controlled by a back pressure regulator (BPR). The precursor and ScH_2O streams were fed into a Separex high pressure CSTR with a capacity of 10 ml through top screw-in tubes. The product slurry came out from the bottom of the reactor, and then cooled down to room temperature by passing through a jacketed cooler. The cooled product slurry was recovered after depressurization by BPR to atmospheric pressure and collected in the product tank.

During the synthesis process, the feed flow rates for both TiBALD solution and distilled water were maintained at 10 ml/min. The system pressure was adjusted by the BPR and kept at 24.1 MPa. The precursor stream concentration is 0.01wt%, which was obtained by diluting the 50wt% purchased TiBALD solution using distilled water. Supercritical water temperature was controlled at 400°C by adjusting oven setting temperature.

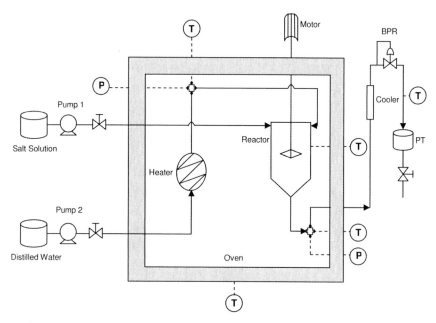

Figure 1   *Experimental apparatus for continuous hydrothermal synthesis of TiO₂ nano-particles*

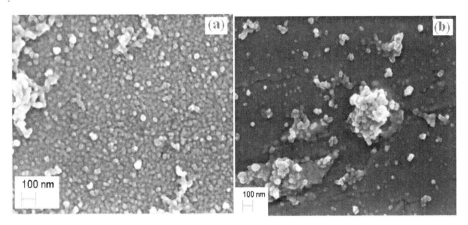

Figure 2   *SEM images of TiO$_2$ nano-particles: (a) primary particles (b) particle aggregates*

4   RESULT AND DISCUSSION

A drop of the TiO$_2$ product slurry taking from the collection tank was placed on SEM standard measurement plate and left for air-drying for 15 min. SEM images were then taken with different magnifications. It is observed from these images that fine spherical shape TiO$_2$ nano-particles were formed. SEM images of particles produced are shown in Figure 2. The images indicate that particle mean diameter is about 20-30 nm. As can be seen in Figure 2b, among the individual particles, some large aggregates were also formed, and the largest aggregate in the image has a size of over 250 nm.

The image analysis software ImageJ[11] was used to analyse the obtained SEM images. By adjusting the threshold levels of the images, the size information for primary particles can be identified. For each TiO$_2$ sample, the SEM images taken at different parts of the sample were analysed and PSDs were calculated based on the size data obtained from images. Figure 3a shows the PSDs computed using a total of 11839 primary particle size results recorded from analysing 24 SEM images.

Figure 4a shows the PB simulation results of PSD of particles produced in the CHS system. Each distribution curves in the figure represents PSDs at a specific residence time. For a CSTR reactor, the residence time is calculated by:

$$\tau = \frac{V}{W} \qquad\qquad (5)$$

As the synthesised products were produced within a 10 ml CSTR reactor and the flow rate was kept at 20 ml/min, the residence time for such a system, according to Eq.(5), is 30 s. The PB modelling result of PSD with 30 s residence time was shown in Figure 4b. The produced particle mean size is about 30 nm, which is consistent with the SEM results of 20-30nm. For the 0-100 nm size group particles, the model gives a similar distribution pattern (Figure 3b), compared with the image analysis results shown in Figure 3a.

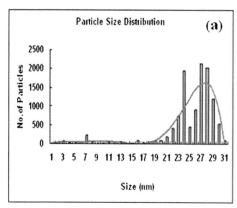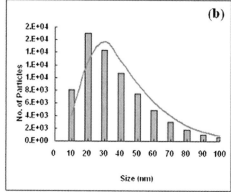

Figure 3   *PSD of CHS system with 30s residence time: (a) experimental result, (b) modelling result*

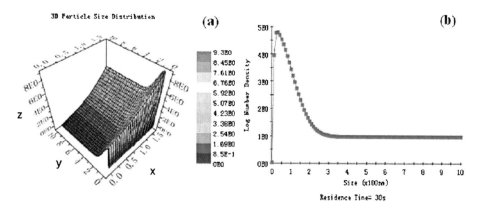

Figure 4   *(a) Simulated PSDs of systems with different residence time (axis: x - residence time (s); y - size (×100 nm); z - log number density), (b)Predicted PSD of CHFS system with 30s residence time*

5   CONCLUSION

A population balance model was developed to predict the size distribution of nano-particles synthesised using a CHS reactor. In order to validate the model, an experiment rig was built to produce nano-size TiO_2 particles in a CSTR reactor. Product particle morphology and size distributions were characterised by SEM and image analysis methods. It was found that the mean diameter of the produced nano-particles is around 20-30 nm. Model validation indicated that the simulation results are in agreement with the experimental data, in particular, the nano-particle mean size.

References

1.   J. A. Darr and M. Poliakoff, *Chem Rev,* 1999, **99**(2), 495-541.

2. S.-I. Kawasaki, Y. Xiuyi, K. Sue, Y. Hakuta, A. Suzuki and K.Arai, *Journal of Supercritical Fluids,* 2009,**50**(3), 276-282.

3. Z. C. Zhang, S. Brown, J. B. M. Goodall, X. L. Weng, K. Thompson, K. N.Gong and J. A. Darr, *Journal of Alloys and Compounds,* 2009, **476**(1-2), 451-456.

4. E. Lester, P. Blood, J. Denyer, D. Giddings, B. Azzopardi and M. Poliakoff, *Journal of Supercritical Fluids,* 2006, **37**(2), 209-214.

5. A. Y. Sheikh, A. G. Jones and P.Graham, *Zeolites,* 1996, **16**(2-3), 164-172.

6. N. Lummen and B. Kvamme, *Journal of Supercritical Fluids,* 2008, **47**(2), 270-280.

7. A. Erriguible, F. Mariasi, F. Cansell and C. Aymonier, *Journal of Supercritical Fluids,* 2009, **48**(1), 79-84.

8. J. Sierra-Pallares, M. Teresa Parra-Santos, J. García-Serna, F. Castro and M. José Cocero, *Journal of Supercritical Fluids,* 2009, **50**(2), 146-154.

9. M. Chen, C. Y. Ma, T. Mahmud, J. A. Darr and X. Z. Wang, *Journal of Supercritical Fluids,* acccepted in July 2011. doi:10.1016/j.supflu.2011.07.002

10. C. B. B. Costa, M. R. W. Maciel and R. Maciel, *Computers & Chemical Engineering,* 2007, **31**(3), 206-218.

11. ImageJ software, http://rsbweb.nih.gov/ij/

SPHERICAL GAMMA ALUMINA NANO-POWDERS PRODUCED FROM ALUMINIUM SALTS BOEHMITE SOL BY SPAY-PYROLYSIS PROCESSES

X. Wu, S Ye and Y. Chen

State key Laboratory of Multiphase Complex System, Institute of Process Engineering of Chinese Academy of Sciences, Beijing, 100190, P.R. China
Email: wxftsjc@mail.ipe.ac.cn or yfchen@mail.ipe.ac.cn.

1 INTRODUCTION

The hot-wall aerosol synthesis, or spray pyrolysis (SP), offers the significant advantages over other material processing techniques as gas-to-particle conversion, liquid or solid-state processing followed by milling in industrial applications. Within this aerosol procedure, the evaporation, precipitation, drying and decomposition of the precursors occur in the droplet reactors in a single step. The resulting particles, or secondary particles, arise from the growth and agglomeration of primary nanoclusters generated within droplet reactors.[1,2] Therefore, it is expected that the properties of droplets and precursors exert the remarkable influences on the final size and distribution and microstructures of the resulting products.

Alumina nanopowders have important applications in manufacturing fine ceramics, heterogeneous catalysis, fillers in metal-matrix composites and so on. Generally, alumina products have preferentially been required with small particle size and narrow size distribution, large surface area, spherical morphology and absence of agglomerates, in order to improve properties of alumina-based materials. The ultrasonic spray-pyrolysis (USP) technique has been considered as a versatile SP process applied in fabricating inorganic powders with spherical morphology and controlled microstructures.[2-5] However, in the USP process, the co-generated hollow and/or foaming porous structures, even collapsing, are usually formed due to surface precipitation and/or poor thermoplastics of solutes within aerosol process of atomized droplets,[2,3] which broaden the particle size distribution and worsening the properties of the final products.

In the present work, SP synthesis of spherical alumina is comparatively investigated using aluminium salt and beohmite sol as the starting materials, respectively. The morphology and structures, crystallinity and thermal behaviours are also examined. The formation process of alumina products derived from different precursors are also explored.

2 EXPERIMENTAL

2.1 Preparation of Spherical Alumina Nanopowders

The aluminium nitrate ($Al(NO_3)_3.9H_2O$, AR) and ammonia is used as received. The used boehmite sols were pre-prepared before USP process as follow: In A typical procedure, the excess amount of ammonia was drop-wise added into 0.3M Al $(NO_3)_3$ solution with vigorous stirring till pH close to 9.5. Then, the generated viscous white slurry was sealed with plastics film and aged statically for 12 hours. After that, the precipitation was filterd and washed with deionized water till pH close to 6.5. The as-obtained precipitation was re-dispersed into deionized water to 120ml suspension (0.25M eqv. Al^{3+}content) in a ultrasonic bath. The 0.5M HNO_3 solution was added drop-wise into the above suspension with molar ratio of H^+/Al^{3+}=0.3. Finally, the white slurry was sealed and transferred into electric oven for preparation of boehmite sol under fixed temperature of 80 °C for 8h. The transparent but slight blue sol was prepared for USP usage. In addition, the 0.25M $Al(NO_3)_3$ solution were also prepared for preparation of spherical alumina in USP process.

The custom-made pilot-scale SP device was similar to that reported by Martin *et al.*[2] The apparatus was comprised of a connected ultrasonic nebulizer, a vertical quartz tube with two-zone heating furnaces and particle capturing filters. The sprayed droplets of precursor solutions were carried by air into quartz tubes at 250~450 °C in the lower zone (drying and thermolysis zone, or DTZ) and 750°C in the upper sintering zone (or SZ), respectively. The generated particles were captured by grass-fibre filters. The air inflow rate is fixed at 20 mL/s.

2.2 Characterization

A scanning electron microscope (SEM) of type JSM-6700F (JEOL) was used to characterize the morphologies of the products. Transmission electron microscopy (TEM) images were taken using a JEM-2010 JEOL instrument at an accelerating voltage of 200KV. The crystallographic structures of the solid samples were determined by the X'Pert Pro powder diffractometer with high-intensity Cu-Kα radiation (λ=1.54187 Å). The TG-DSC analysis was conducted on a NETZSCH thermal analyser (STA 449).

3 RESULTS AND DISCUSSION

The morphologies of the USP samples produced with boehmite sol and/or aluminium nitrate solution are shown in Figure 1, respectively. It is found that the produced particles are essentially sperical. More concretely, in contrast with the USP samples in Figures 1D-1F from $Al(NO_3)_3$ solution, rough surfaces are observed for those from alumina samples using boehmite sol as shown in Figures 1A-1C. It can also be seen that the particle size of the samples produced with boehmite sol are smaller than those with equivalent molar $Al(NO_3)_3$ solution under the identical synthesis proceudres. Especially, when the DZT and SZ temperature are set at 350 °C and 750 °C, the average size of as-prepared alumina powder is ca.450 nm with a narrow size distribution, while the USP sample of $Al(NO_3)_3$ solution has an average diameter of 1.25 µm with a wide size distribution (0.2~2 µm). Furthermore, when the DTZ increased to 450 °C, the broken hollow or porous particles with wider size distributions (0.25~70 μm) are identified for the USP samples produced with $Al(NO_3)_3$ solution, as shown in Figure 1F. In contrast, the samples produced using boehmite sol in Figure 1C have an average particle diameter of ca.780 nm with a wide size distribution. Comparatively, it can be observed that, as the DTZ temperature increased from 250 °C to 450 °C, the particle size distribution for the samples produced with boehmite sol was more uniform at a DTZ temperature of 350 °C, but a wider size distribution was obtained at 450 °C, as shown in Figure 1A-1C. while for the samples produced using $Al(NO_3)_3$ solution, the particle size distribution became increasingly wide as shown in Figure 1D-1F. It can be concluded that, with equivalent molar Al element contents and under identical synthesis conditions, the boehmite-derived USP process produces smaller particles with a narrower size distribution than the aluminium salt-derived USP products.

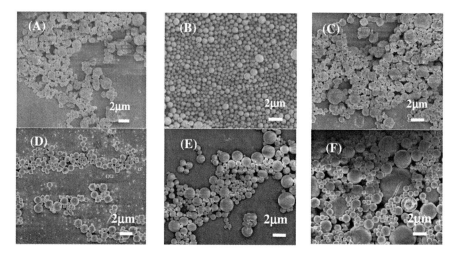

Figure 1 *SEM images of as-prepared spherical alumina by USP method using boehmite sol (A-C) and $Al(NO_3)_3$ solutions(D-F) under DTZ temperature of 250 °C (A,D), 350 °C (B, E) and 450 °C (C,F).*

The microstructures of USP samples using boehmite sol and/or equivalent molar $Al(NO_3)_3$ solution are shown in Figure 2. As shown in Figure 2A, the samples produced with boehmite sol possess solid structure even though the particles are not perfectly spherical. This is further conformed from the TEM image (the large inset) also confirm solid structure, and the HRTEM images (the small inset) reveals that the individual solid particles are actually agglomerates with nanocrystals of ca.5nm. However, only smooth surface and broken hollow structure with larger diameter are clearly identified for the USP samples produced using $Al(NO_3)_3$ solution, as shown in Figure 2B. Accordingly, it can be expected that, during the USP process, the sprayed droplets from boehmite sol and/or $Al(NO_3)_3$ solution experience different precipitation, particle nucleation and growth processes. It is well known that, in the salt solution-derived USP process, the surface precipitation of solutes within sprayed droplets usually occurred, which template the spherical droplets at the gas-liquid interface and produce hollow or porous structure with solvent evaporation continuously.[1,3] For USP alumina powders produced with $Al(NO_3)_3$ solution (Figure 2B), they were very likely produced through the commonly evaporation-precipitation-pyrolysis (or EPP) process.[1] The solid structure of the USP alumina powder produced with boehmite sol (Figure 2A) indicates that the evaporation-aggregation-pyrolysis(EAP) process occurs, during which the boehmite nanocrystals aggregated into spherical blocks, as the solvent was evaporating, and then thermolyzed into alumina.

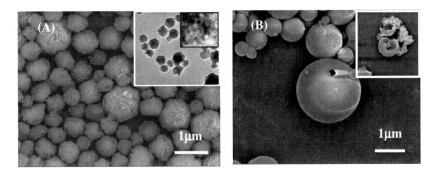

Figure 2 *SEM and/or TEM images (insets in A) of boehmite-derived spherical alumina nanopowders (A) and SEM images of $Al(NO_3)_3$-derived spherical alumina(B)*

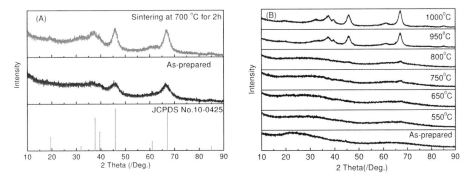

Figure 3 *XRD patterns of as-prepared spherical alumina and post-sintered alumina using boehmite sol (A) and $Al(NO_3)_3$ solution (B)*

The crystallinity of both USP alumina products are investigated, as shown in Figure 3. It is found from Figure 3A that, the as-prepared samples using boehmite sol have γ type crystal phase (JCPDS No. 10-0425) with slightly improved crystallinity when post-sintering at 700 °C for 2h. While, the as-prepared sample using $Al(NO_3)_3$ solution possesses amorphous nature, and the γ type crystal phase transformation from amorphous one occurs above 800 °C. Obviously, the boehmite-derived USP process for preparing alumina powders is an energy-saving process. The thermal performances of both as-prepared alumina powders are shown in Figure 4A and 4B, respectively. It is observed in Figure 4A that, at a temperature of 30~200 °C, the mass loss (ca.13%) and weak endothermic peak are identified, which is attributed to dehydration of physisorbed water.

Then, a second mass loss (ca.17%) and slightly high exothermic peak centred at 413.6 °C are observed, which is attributed to dehydration of chemisorbed water and the crystal transformation of boehmite into γ–Al_2O_3.[6] At a temperature of 450~1000°C, no obvious mass loss is observed, and a strong endothermic peak is attributed to surface dehydration.[6] While for the USP samples produced with $Al(NO_3)_3$ solution, no obvious mass loss and endothermic peak are observed at the temperature range of 200~792 °C, besides the dehydration of physisorbed water at a low temperature range of 30~200 °C, and then a significant mass loss and sharp endothermic process centred at 829.5 °C are identified, which is attributed to the transformation of amorphous $Al(OH)_3$ into γ–Al_2O_3.[7] It is concluded that the as-prepared alumina using $Al(NO_3)_3$ solution is amorphous $Al(OH)_3$ phase, which is consistent with the XRD result and different from γ–Al_2O_3 produced with the boehmite-derived USP process.

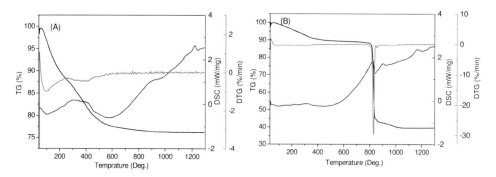

Figure 4 *Thermal analysis of as-prepared spherical alumina nanopowders using boehmite sol (A) and Al(NO$_3$)$_3$ solution (B).*

4 CONCLUSIONS

Using a pilot-scale spay-pyrolysis (SP) device, spherical Alumina powders have been fabricated using boehmite sol and Al (NO$_3$)$_3$ solutions, respectively. It is found that the salt-derived alumina powders have a broad diameter range (0.2~70 μm) and hollow structures with amorphous phase. In comparison, the alumina powders using boehmite sol under the fixed SP conditions consist of solid particles with average diameter of 0.45 μm with a narrow size distribution and gamma crystal phase. Furthermore, the boehmite-derived alumina is formed with aggregates made of less 5 nm nanocrystals. The difference in size and microstructures is attributed to the evaporation-precipitation-pyrolysis (EPP) mechanism for Al (NO$_3$)$_3$-derived process and the evaporation-aggregation-pyrolysis (EAP) for boehmite -derived process, respectively.

Acknowledgements

This research is supported by National Natural Science Foundation of China (NSFC) (No. 51002154).

References

1. G.L. Messing, S.C.Zhang, G.V. Jayanthi, *J. Am. Ceram. Soc.* 1993, **76**, 2707.
2. M. I. Martin, L.S.Gomez, O. Milosevic, M.E.Rabanal, *Ceram. Inter.* 2010, **36**,767.
3. W. N. Wang, A. Purwanto, I. W. Lenggoro, K. Okuyama, H. Chang, H. D. Jang, *Ind. Eng. Chem. Res.,* 2008, 47, 1650.
4. S. E. Skrabalak, K. S. Suslick, *J. Am. Chem. Soc.*, 2006,128, 12642.
5. J.M.A. Caiut, J. Dexpert-Ghys, Y. Verelst, H. Dexpert, S.J.L. Ribeiro, Y. Messaddeq, *Powder Tech.,* 2009,190, 95.
6. M. Nguefack, A. F. Popa, S. Possignol, C. Kappenstein, *Phys. Chem. Chem. Phys.* 2003, **5**, 4279.
7. J. Y. Park, S. G. Oh, U. Paik, S. K. Moon, Mater. Lett., 2002, 56, 429.

A FACILE ROUTE FOR RAPID SYNTHESIS OF MESOPOROUS SBA-16 SILICA

S. W. Xu,[1] H. P. Pu,[1] H. Wang,[2] L. Y. Zhang,[1] C. Y. Han,[1] D. Q. Du,[1] Y. Y. Zhang[1] and Y. M. Luo[1,2]

[1] Faculty of Environmental Science and Engineering, Kunming University of Science and Technology, Kunming 650093, China. E-mail: luoyongming@tsinghua.org.cn.
[2] Faculty of Metallurgical and Energy Engineering, Kunming University of Science and Technology, Kunming 650093, China.

1 INTRODUCTION

SBA-16 has attracted a lot of attention for potential applications in adsorption, catalysis, and separation as well as in immobilization of biomolecules and catalysts because of its 3D channel systems and uniform-sized pores of superlarge cage-like structures,[1,2] which are less susceptible to pore blockage and is favorable for the transport of reactants and products.[3] In general, SBA-16 has been synthesize by using triblock copolymers F127 or F108 as a template under strongly acidic conditions (pH<1.0), which proceeds through a $(S^0H^+)(X^-I^+)$ pathway.[1,4-6] It is obvious that such a strongly acidic medium is disadvantageous for the industrialization of SBA-16 due to some practical problems such as eroding the vessels, being harm to workers and causing other environmental problems.[7] Nevertheless, reducing molar ratio of liquid acids to Si would cause the crystallization time to be substantially increased. Stucky et al. [8] reported that the crystallization time required for synthesis of ordered mesoporous silica materials with conventional method would increase from 24 h to 72 h if the molar of HCl to Si is lower than 0.59 and without adding any promoter, which serves as a perfect example. More importantly, weakly acidic media is prone to the formation of amorphous or otherwise disordered silica because of the absence of sufficiently strong electrostatic or hydrogen-bonding interactions.[9] On the other hand, it is well-known that most molecular sieves synthesized by hydrothermal methods, which generally suffer from the drawback of long crystallization time.[10] Although the crystallization time for microwave synthesis of SBA-16 can be shortened to within 2h, the molar ratio of HCl to Si is as high as 6.68.[11] Therefore, it can be concluded that reducing HCl concentration to large extent, substantially shortening crystallization time, and whilst maintaining the well-ordered structure is a challengeable task.

Recently, limited synthesis efforts have been made to prepared SBA-16 under weak acidic media, involving the addition of cationic fluorocarbon surfactant,[12] the use of two-step route,[7] the use of co-surfactants [13,14] and / or ternary surfactants.[15,16] Despite significant progress in the synthesis of SBA-16 under low concentration of HCl media, the corresponding aged time is as long as 24 h or even more. Therefore, further studies in this area are desirable, especially in relation to rapid synthesis, thermal and hydrothermal stability.

2 EXPERIMENTAL

Tetraethylorthosilicate (TEOS) and F127 were employed as a silica source and a structure-directing and to synthesize ordered mesoporous silica materials described in this article, and the corresponding procedures were followed by our previous report for the synthesis of ordered mesoporous SBA-15[17] except for adding inorganic salts (NaCl, KCl, Na_2SO_4 or K_2SO_4) and adjusting reaction temperature and the starting chemical composition. The samples synthesized with adding NaCl and aged at 100°C for 3h, 6h, 12h and 24h were designated as MSM-1, MSM-2, MSM-3 and MSM-4.

The hydrothermal stability was investigated by mixing ca. 0.5 g of calcined MSM-4 with 100 g deionised water and heating in a closed bottle at 100 °C under static conditions.

3 RESULTS AND DISCUSSION

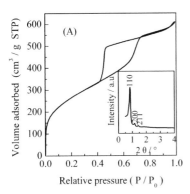 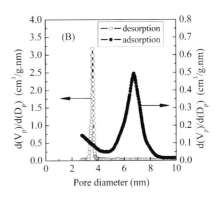

Figure 1 *N_2 adsorption-desorption isotherm and XRD pattern (inset) and (B) BJH Pore size distribution curves of MSM-4.*

Table 1 *Physicochemical properties of mesoporous silica materials synthesized under weak acidic conditions.*

Sample	HCl/Si[a]	CT[b] (h)	a_0[c] (nm)	S_{BET} (m^2g^{-1})	TPV[d] (cm^3g^{-1})	MPV[e] (cm^3g^{-1})	Pore diameter (nm) CSD[f]	ESD[g]	CS[h]
MSM-1	0.21	3	–[i]	651	0.76	0.52	–[i]	–[i]	DIS[j]
MSM-2	0.21	6	16.1	713	0.82	0.59	6.5	3.3	Cubic
MSM-3	0.21	12	16.4	745	0.86	0.64	6.7	3.5	Cubic
MSM-4	0.21	24	16.6	776	0.90	0.69	6.9	3.6	Cubic
MSM-4[k]	–[i]	–[i]	16.5	629	0.91	0.72	7.5	4.0	Cubic
MSM-C[l]	5.85	24	16.5	735	0.83	0.59	6.6	3.4	Cubic
MSM-C[m]	–[i]	–[i]	–[i]	316	0.55	0.34	–[i]	–[i]	DIS[j]

[a] Molar ratio. [b]CT = Crystallization time. [c]unit cell parameter, $a_0 = d_{110} \times \sqrt{2}$. [d]TPV = Total pore volume. [e]MPV= Mesoporous pore volume. [f]The cage size distribution calculated from the adsorption branch. [g]The entrance size distribution calculated from desorption branch. [h]CS = Crystal structure was characterized by XRD. [i]No data. [j]DIS = disordered. [k]The

sample MSM-4 was treated with 100 °C water for 120 h. [l]Sample MSM-C was prepared by refence 1. [m]The sample MSM-C was treated with 100 °C water for 120 h.

A type IV isotherm with a H_2-type hysteresis loop starting at a relative pressure of 0.6-0.7 and abruptly ending at 0.45 are observed for the sample MSM-4, indicating the presence of good-quality cage-like pores.[18,19] Three well-resolved diffraction peaks in the range of 2θ blow 1.5°, indexed as (110), (200) and (211) reflections, are detected in the XRD pattern of MSM-4, which corresponds to a well-ordered body-centered cubic (*Im3m*) mesostructure. The entrance and the corresponding cage size of MSM-4 calculated from the desorption and adsorption branch of the isotherm are 3.6 and 6.8 nm, respectively. The cell parameters, BET surface area and total pore volume are 16.6 nm, 776 m^2g^{-1} and 0.90 cm^3g^{-1}, which is comparable to those of MSM-C synthesized with conventional synthesis route.[1,20]

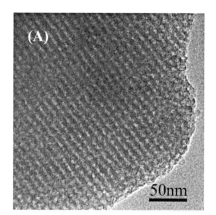 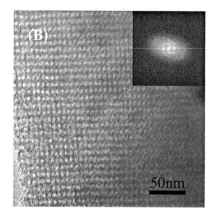

Figure 2   *TEM images of MSM-4*

A well-ordered mesostructure is observed for MSM-4, and uniform pore size distribution along the (100) and (111) direction[5,21] appears to be continuous in range of several hundred nanometers (Figure 2). Combining the result of XRD characterization which show that three diffraction peaks at 2θ values between 0.5 and 1.5 °, assigned to the (110), (200) and (211) reflections, are observed for MSM-4. Therefore, it should be concluded that a well-ordered body-centered mesostructure with the *Im3m* space group was formed within MSM-4.

MSM-2, MSM-3 and MS-4 exhibit type IV isotherms, and a H_2-type hysteresis loop is detected for each of them. It is well-demonstrated that the position of the inflection point is clearly related to a diameter in the mesopore range, and the sharpness of these steps indicates the uniformity of the mesopore size distribution.[21,22] As for SBA-16-type silica materials, the entrance size and the cage size were calculated by the use of the data from the desorption and the adsorption branch, respectively.[5,23,24] It is noticeable that the inflection points of the adsorption and desorption isotherms slightly shift to high relative pressure with crystallization time, which might assigned to the increase both in the pore size and cage size. This change can be explained as follows: it is well-known that $(EO)_m(PO)_n(EO)_m$ triblock copolymers can form micelles in water, where the $(PO)_n$ moieties exhibit more strong hydrophobicity than the $(EO)_m$ moieties.[25] Moreover, the hydrophobicity of the $(EO)_m$ moieties increases with temperature.[26,27] The long crystallization time is expected to be in favour of the hydrophobicity of the $(EO)_m$ moieties, thus resulting in the hydrophobic domain volumes and the length of $(PO)_n$ moieties increasing, corresponding to the reduce of the hydrophilic domain volumes and the occlusion of the $(EO)_m$ moieties in the as-synthesized matrix, which are responsible for d_{110} spacing, pore size and pore volume increasing. In addition, it is also noted that the hysteresis loop of MSM-1 is different from that of MSM-2, MSM-3 and MSM-4, and the difference might be ascribed to the formation of otherwise disordered or amorphous mesoporous silica materials, as further confirmed by XRD. Figure 3B provides small angle XRD patterns of MSM-1, MSM-2, MSM-3 and MSM-4. As shown in Figure 3B, the presence of a single diffraction peak in 2θ region blow 2° for MSM-1 is indicative of a disordered mesostructure.[28-30] Therefore, it can be concluded that the MSM-1 is only a mesostructure rather than an ordered mesostructure due to the absence of other well-resolved diffraction peaks. Obviously, three well-resolved peaks in the region of 0.8-1.5° indexed to the (110), (200) and (211) reflections are observed for MSM-2, MSM-3 and MSM-4, which can be associated with the body-centered cubic space group (*Im3m*). These results indicate that the SBA-16-type mesoporous silica materials can be synthesized within 6 h, which is very short compared to the crystallization time required for other synthesis method.[1,7,14,16]

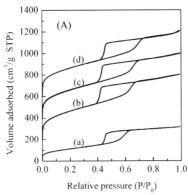 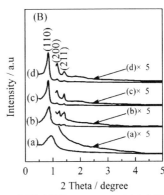

Figure 3   *(A) N_2 adsorption-desorption isotherms and (B) XRD patterns of (a) MS-1, (b) MS-2, (c) MS-3 and (d) MS-4. The isotherms of (b), (c) and (d) are offset by 200, 400 and 600 cm^3g^{-1} for clarity, respectively.*

A type IV isotherm along with a type-H_2 hysteresis loop is still observed for MSM-4 after treated with treated with 100 °C water for 120 h, while the jumps of the adsorption and desorption branches slightly shift to high relative pressure, implying both the entrance and the cage size of the sample increase. Furthermore, the presence of three diffraction peaks in the XRD pattern of MSM-4 after hydrothermal treatment is an indication of a good *Im3m* cubic ordering. In addition, BET surface area and total pore volume of MSM-4 after hydrothermal stability test are 629 m^2g^{-1} and 0.91 cm^3g^{-1}, suggesting MSM-4 maintains more than 81% and 100 % of its original surface area and pore volume. In contrast, MSM-C only retains about 43% and 66% of its BET surface area and total pore volume. More importantly, ordered mesostructure of MSM-C was completely destroyed after the same hydrothermal treatment. In light of these results, it can be concluded that MSM-4 exhibits far higher hydrothermal stability than MSM-C, which might be attributed to the higher degree of polymerization and condensation of silanol groups that originated from the synergistic effect of rapid nucleation with the aid of HPMo and salt effect during the crystallization process.[17,31]

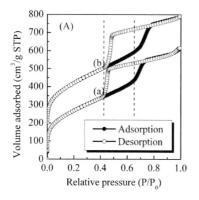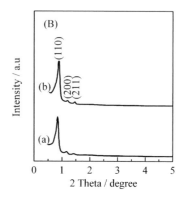

Figure 4   *(A)N_2 adsorption-desorption isotherms and (B) XRD patterns of MSM-4 before and after hydrothermal stability test: (a) before and (b) after treated with 100 °C water for 120 h. The isotherm of (b) is offset by 200 $cm^3 g^{-1}$ for clarity.*

4 CONCLUSIONS

In summary, mesoporous silica materials (MSM) with highly cubic ordered mesostructure (*Im3m*) have been synthesized within 6h under weak acidic media by adding small amount of phosphomolybdic acid and inorganic salts. The reaction products have still maintained a well-ordered body-centered cubic (*Im3m*) mesostructure even after treated with 100 °C water for 120 h. High hydrothermal stability of them might be attributed to the higher degree of polymerization and condensation of silanol groups, originated from the synergistic effect of rapid nucleation with the aid of HPMo and salt effect during the crystallization process.

Acknowledgements

This research work was supported by National Natural Science Foundation of China (Grant No. 20867003 and 51068010), Postdoctoral Science Foundation Funded Project (Grant No. 20100471686), Young Academic and Technical Leader Raising Foundation of Yunnan Province (Grant No. 2008py010).

References

1   D. Y. Zhao, Q. H. Huo, J. L. Feng, B. F. Chmelka and G. D. Stucky, *J. Am. Chem. Soc.*, 1998, **120**, 6024.
2   Y. Sakamoto, M. Keneda, O. Teresaki, D. Y. Zhao, J. M. Kim, G. D. Stucky, H. J. Shin and R. Ryoo, *Nature*, 2000, **408**, 449.
3   R. M. Grudzien, B. E. Grabicka, M. Jaroniec, *J. Mater. Chem.*, 2006, **16**, 819.
4   H. Wang, C. Y. Han, H. P. Pu, L. Y. Zhang, Z. H. Zou and Y. M. Luo, *Bull. Chem. Soc. Jpn.*, 2010, **83**, 852.

5   X. Meng, D. Lu and T. Tatsumi, *Micropor. Mesopor. Mater.*, 2007, **105**, 15.
6   W. J. J. Stevens, M. Mertens, S. Mullens, I. Thijs, G. V. Tendeloo, P. Cool and E. F. Vansant, *Micropor. Mesopor. Mater.*, 2006, **93**, 119.
7   Z. W. Jin, X. D. Wang and X. G. Cui, *J. Colloid Interface Sci.*, 2007, **307**, 158.
8   P. Schmidt-Winkel, P. D. Yang, D. I. Margolese, B. F. Chmelka and G. D. Stucky, *Adv. Mater.*, 1999, **11**, 303.
9   D. Zhao, J. Feng, Q. Huo, N. Melosh, G. Fredrickson, B. F. Chmelka and G. D. Stucky, *Science*, 1998, **279**, 548.
10 R. Kumar, A. Bhaumik, R. K. Ahedi and S. Ganapathy, *Nature*, 1996, **381**, 298.
11 Y. K. Hwang, J.-S. Chang, Y.-U. Kwon and S.-E. Park, *Micropor. Mesopor. Mater.*, 2004, **68**, 21.
12 Y. Han and J. Y. Ying, *Angew. Chem. Int. Ed.*, 2005, **44**, 288.
13 D. F. Li, X. Y. Guan, J. W. Song, Y. Di, D. X. Zhang, X. Ge, L. Zhao and F. S. Xiao, *Colloids and Surfaces A: Physicochem. Eng. Aspects.*, 2006, **272**, 194.
14 M. Mesa, L. Sierra and J. Patarin, J. L. Guth, *Solid State Sci.*, 2005, **7**, 990.
15 B. C. Chen, M. C. Chao, H. P. Lin and C. Y. Mou, *Micropor. Mesopor. Mater.*, 2005, **81**, 241.
16 C. L. Lin, Y. S. Pang, M. C. Chao, B. C. Chen, H. P. Lin, C. Y. Tang and C. Y. Lin, *J. Phys. Chem. Solid.*, 2008, **69**, 415.
17 Y. M. Luo, Z. Y. Hou, R. T. Li and X. M. Zheng, *Micropor. Mesopor. Mater.*,2008, **109**, 585.
18 J. R. Matos, M. Kruk, L. P. Mercuri, M. Jaroniec, L. Zhao, T. Kamiyama, O. Terasaki, T. J. Pinnavaia and Y. Liu, *J. Am. Chem. Soc.*, 2003, **125**, 821.
19 R. M. Grudzien, B. E. Grabicka and M. Jaroniec, *Appl. Surf. Sci.* 2007, **253**, 5660.
20 B. R. Jermy, S. Y. Kim, K. V. Bineesh, D. W. Park, *Micropor. Mesopor. Mater.*, **2009**, 117, 661
21 B. L. Newalkar, S. Komarneni and H. Katsukib, *Chem. Commun.*, 2000, 2389.
22 Y. M. Luo, Z. Y. Hou, R. T. Li and X. M. Zheng, *Micropor. Mesopor. Mater.*, 2008, **110**, 583.
23 R. Rockmann and G. Kalies, *J. Colloid. Interf. Sci.*, 2007, **315**, 1.
24 H. Sun, Q. H. Tang, Y. Du, X. B. Liu, Y. Chen and Y. H. Yang, *J. Colloid. Interf. Sci.*, 2009, **333**, 317.
25 M. Impéror-Clerc, P. Davidson and A. Davidson, *J. Am. Chem. Soc.*, 2000, **122**, 11925.
26 M. Kruk, M. Jaroniec and C. H. Ko, R. Ryoo, *Chem. Mater.*, 2000, **12**, 1961.
27 G. Wanka, H. Hoffmann and W. Ulbricht, *Macromolecules*, 1994, **27**, 4145.
28 P. T. Tanev and T. J. Pinnavaia, *Science*, 1995, **267**, 865.
29 S. A. Bagshaw, E. Prouzet and T. J. Pinnavaia, *Science*, 1995, **269**, 1242.
30 Y. M. Luo, Z. Y. Hou, J. Gao, D. F. Jin and X. M. Zheng, *Mater. Lett.*, 2006, **60**, 393.
31 R. Ryoo, S. Jun, *J. Phys. Chem. B.*, 1997, **101**, 317.

THE UNDERLYING MECHANISM OF HYDROTHERMAL STABILITY FOR ORDERED MESOPOROUS MOLECULAR SIEVES WITHOUT MICROPORES

H. P. Pu,[1] H. Wang,[2] S. X. Xu,[1] C. Y. Han[1], L. Y. Zhang[1], Y. Y. Zhang[1], D. Q. Du[1], Y. M. Luo[1,2]

[1]Faculty of Environmental Science and Engineering, Kunming University of Science and Technology, Kunming 650093, P.R. China. E-mail: luoyongming@tsinghua.org.cn.
[2] Faculty of Metallurgical and Energy Engineering, Kunming University of Science and Technology, Kunming 650093, China.

1 INTRODUCTION

Micropore-free SBA-15 is a model porous solid for investigating the behaviour of matter in confined space and using as a host for the fabrication of nanowires and related materials in nanotechnology.[1,2] In the past two decades, much effort has been devoted to synthesize mesoporous SBA-15 materials by manipulating the synthesis conditions,[3-6] which involve adding inorganic ionic salts,[1,2,6] adjusting the ratio of surfactant to TEOS,[3] varying synthesis temperature,[3,4] employing co-templates,[4,5] and using microwave hydrothermal synthesis[6] as well as post-treatment.[7] However, the hydrothermal stability of mesoporous SBA-15 materials synthesized with the above methods is relatively low due to synthesis route's limitation. For instance, Pinnavaia and co-workers pointed out that the use of high concentrations of salts may lead to salt occlusion in the crystallized framework which could impart poor hydrothermal stability in the framework.[8] Therefore, it is of importance to design a new route to achieve the control over framework microporosity of SBA-15 type mesoporous materials without compromising hydrothermal stability.[9]

In this study, we developed an effective route to synthesize two ordered mesoporous molecular sieves (MSAMA-2 and MSAMS-4). Not only these two ordered mesoporous molecular sieves are micropore-free but also they show extraordinary stability both in boiling water and in high temperature steam, which might be ascribed to the formation of five-membered ring subunits within their framework pores.

2 EXPERIMENTAL

MSAMS-2 and MSAMS-4 have been synthesized by following procedures, which are mainly composed of three steps: the first step is synthesis of SBA-15, and the second and the third step consist of preparation of the diluted solution of aluminosilicate sol-gel and re-crystallization of SBA-15 both within the diluted solution of aluminosilicate sol-gel and glycerol. In the first step, SBA-15 was synthesized using P123 as a structure-directing agent, and tetraethyl orthosilicate (TEOS) as a silica source. The starting chemical composition of the reaction mixture was 1 mol P123: 60 mol TEOS: 390 mol HCl: 8536 mol H_2O. Then, the mixture was stirred at 38 °C for 20 h and aged at 96 °C under static conditions for 24 h, respectively. Subsequently, the reaction products were filtered, washed and dried at 45 °C for 48 h. Finally, the resulting white solid was calcined at 550 °C in air for 8 h. In the second step, tetrapropylammonium (or tetraethylammonium) bromide containing 9.0 mmol of tetrapropylammonium cation (TPA$^+$) or tetraethylammonium cation (TEA$^+$) was dissolved in 9.6 ml of deionized water to yield a solution. Thereafter, 21.6 mmol of TEOS, 25 ml of H_2O and 0.06-0.31 g of sodium aluminate (41.1 wt%Al_2O_3 and 25.9 wt% Na_2O) were added into the solution, and the resulting mixture was stirred at room temperature (RT) overnight. Then, the mixture was diluted with deionized water to yield 150 ml of aluminosilicate sol-gel diluted solution. In the third step, 5 g of calcined SBA-15 was added into 150 ml of the aluminosilicate sol-gel diluted solution under vigorous stirring at RT for 2 h. After that, the reaction product was crystallized at 110-130 °C for 12-36 h, followed by filtered, washed and dried at 80 °C for 24 h. Subsequently, the resulting solid was suspended in 12.5 ml of glycerol and crystallized at 130-150°C for 24-48 h. At last, the product was filtered, washed, dried, and calcined at 550 °C in air for 8 h.

The hydrothermal stability was examined by steamed the samples with 800 °C vapour (20 vol% H_2O in N_2) for 6 h or mixed ca. 0.5 g of sample with 50 ml of deionised water and heating in a closed bottle at 100 °C under static conditions for 360 h.

3 RESULTS AND DISCUSSION

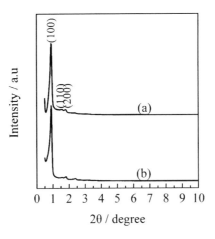

Figure 1   *XRD patterns of (a) MSAMS-2 and (b) MSAMS-4*

As shown in Figure 1, an intense together with two well-resolved diffraction peaks in 2θ values between 0.5 ° and 2 °, indexed to (100), (110) and (200) reflections, are observed for MSAMS-2 and MSAMS-4, which is an indication of a good long-range hexagonal ordering. In addition, it is also noticeable that no diffraction peak in 2θ region of 5 °-10 ° is detected for the two molecular sieves, indicating that MSAMS-2 and MSAMS-4 are of pure mesoporous phases without any bulky zeolite (ZSM-5 and Beta) crystals.[10-14]

In general, the isotherms have been used to estimate the micropore volume within ordered mesoporous materials, which involves several analysis methods such as *t*-plots,[4,6] α_s-plots[15,16] and β-plots.[17] *t*-plots of N_2 adsorption isotherms of MSAMS-2 and MSAMS-4 are illustrated in Figure 2. It is clear that not only the *t*-plots of N_2 adsorption isotherm of MSAMS-2 and MSAMS-4 exhibits a straight line at t = 0.40 - 0.75 nm (t = thickness of adsorbed layer), but also both the extrapolation lines of them go through the origin, thus demonstrating that the MSAMS-2 and MSAMS-4 are free of micropores within their mesoporous framework wall.[3,7]

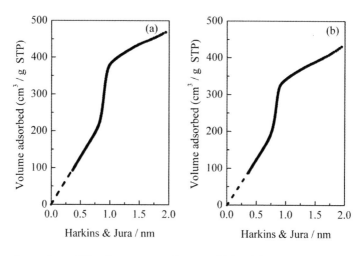

Figure 2   *t-plots of N_2 adsorption isotherms of (a) MSAMS-2 and (b) MSAMS-4.*

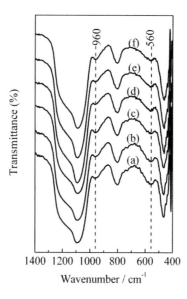

Figure 3   *FT-IR spectra of MSAMS-2 (a, b, c) and MSAMS-4 (d, e, f) before and after hydrothermal stability test. (a) and (d) before, (b) and (e) after treated in 100 °C water for 360 h, (c) and (f) after steamed at 800 °C for 6 h (20 vol % H_2O in N_2).*

It is noticeable that the band centered in the range of 550-600 cm^{-1}, ascribed to characteristic vibration of five-membered ring subunits,[9,18,19] is well-expressed in the spectra of MSAMS-2 and MSAMS-4, which indicates that zeolite-like subunits were formed within these two mesoporous molecular sieves.[20,21] Combining the XRD and *t*-plots of N$_2$ adsorption isotherm characterization results that no diffraction peak in 2θ value of 5 - 10 ° is observed for MSAMS-2 and MSAMS-4 and they are free of micropores, it could be concluded that the zeolite-like subunits with small size other than zeolite crystals were formed within both micropores and mesopores of MSAMS-2 and MSAMS-4. As can be seen in Figure 3 that, even in cases that MSAMS-2 and MSAMS-4 were treated with 100 °C water for 360 h or steamed at 800 °C for 6 h, hardly any changes in the bands both assigned to characteristic vibration of five-membered ring subunits and characteristic vibration of non-condensed silanol groups (centered at 960 cm^{-1}) [22-24] are observed for MSAMS-2 and MSAMS-4. These results demonstrate that the formed five-membered ring subunits are extraordinarily stable, and which should be responsible for improving their hydrothermal stability.[9,18,19]

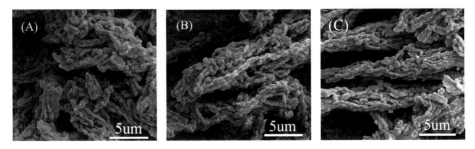

Figure 4   *FE-SEM images of (A) SBA-15, (B) MSAMS-2 and (C) MSAMS-4.*

Field emission scanning electron microscope was used to characterize the morphologies of parent SBA-15, MSAMS-2 and MSAMS-4, and the corresponding FE-SEM images are shown in Figure 4. The morphologies of MSAMS-2 and MSAMS-4 are similar to those of SBA-15. Moreover, it is obvious that neither isolated sphere particles nor isolated cuboid particles of nanoscale, assigned to characteristic morphologies of ZSM-5 and Beta zeolites,[25,26] are observed for MSAMS-2 and MSAMS-4. These results indicate that no bulky ZSM-5 and Beta zeolite crystals were formed within MSAMS-2 and MSAMS-4, which is consistent with XRD characterization results that no diffraction peaks in 2θ values of 5 - 10 ° are detected.

A well-ordered hexagonal mesoporous structure arranged in range of several hundred nanometers is still observed for MSAMS-2 and MSAMS-4 (Figures 5B and 5C). However, the ordering degree of MSAMS-2 and MSAMS-4 is slightly lower than SBA-15. The most likely explanation for the change in ordering degree of MSAMS-2 and MSAMS-4 observed here is that the structural backbone of the parent SBA-15 was well maintained in spite of removing mesoporous structure template P123 during the subsequent re-crystallization within both the diluted solution of aluminosilicate sol-gel and glycerol, while the zeolite-like five-membered ring subunits with small size overgrew within the mesopore channels of SBA-15 with the aid of microporous zeolites structure template TPA$^+$ and TEA$^+$. HR-TEM images, especially the HR-TEM image of MSAMS-2, provide a conceivable evidence to confirm that the zeolite-like five-membered ring subunits with diameter from several angstroms to nanometers are embedded within the channels of MSAMS-2 and MSAMS-4.

Figure 5   *TEM and HR-TEM (insert) images of (A) SBA-15, (B) MSAMS-2 and (C) MSAMS-4*

3 CONCLUSIONS

MSAMS-2 and MSAMS-4 without micropores show extraordinary stability both in boiling water (100 °C over 360 h) and high temperature steam (800 °C for 6 h). The *t*-plots of N_2 adsorptions of them show that these two sieves are free of micropores within their mesoporous framework wall. FT-IR together with XRD, FE-SEM and HR-TEM characterizations demonstrate that the five-membered ring subunits with small size other than zeolite crystals were formed within the framework pores of MSAMA-2 and MSAMS-4. It can be concluded that these zeolite-like subunits are extraordinarily stable, which should play a crucial role in improving the hydrothermal stability for MSAMA-2 and MSAMS-4.

Acknowledgements

This research work was supported by National Natural Science Foundation of China (Grant No. 20867003 and 51068010), China Postdoctoral Science Foundation Funded Project (Grant No. 20100471686), Young Academic and Technical Leader Raising Foundation of Yunnan Province (Grant No. 2008py010).

References

1   J. R. Matos, L. P. Mercuri, M. Kruk and M. Jaroniec, *Chem. Mater.*, 2001, **13**, 1726.
2   B. L. Newalkar and S. Komarneni, *Chem. Mater.*, 2001, **13**, 4573.
3   K. Miyazawa and S Inagaki, *Chem. Commun.*, 2000, **21**, 2121.
4   Y. Han, D. Li, L. Zhao, J. Song, X. Yang, N. Li, Y. Di, C. Li, S. Wu, X. Xu, X. Meng, K. Lin and F.-S. Xiao, *Angew. Chem. Int. Ed.*, 2003, **42**, 3633.
5   D. F. Li, X. Y. Guan, J. W. Song, Y. Di, D. L. Zhang, X. Ge, L. Zhao and F.-S. Xiao, *Colloids and Surfaces A: Physicochem. Eng. Aspects*, 2006, **272**, 194.
6   B. L. Newalkar and S. Komarneni, *Chem. Commun.*, 2002, **16**, 1774.

7   R. Ryoo, C. H. Ko, M Kruk, V. Antochshuk and M. Jaroniec, *J. Phys. Chem. B.*, 2000, **104**, 11465.
8   T. R. Pauly, V. Petkov, Y. Liu, S. J. L. Billinge and T. J. Pinnavaia, *J. Am. Chem. Soc.*, 2002, **124**, 97.
9   Y. M. Luo, Z.Y. Hou, R. T. Li and X. M. Zheng, *Micropor. Mesopor. Mater.*, 2008, **110**, 583.
10 W. Chaikittisilp, T. Yokoi and T. Okubo, *Micropor. Mesopor. Mater.*, 2008, **116**, 188.
11 R. M. Mohamed, H. M. Aly, M. F. El-Shahat and I. A. Ibrahim, *Micropor. Mesopor. Mater.*, 2005, **79**, 7.
12 P. Konova, K. Arve, F. Klingstedtb, P. Nikolov, A. Naydenov, N. Kumar, D. Yu. Murzin, *Appl. Catal. B.*, 2007, **70**, 138.
13 H. Jon, B. Lu, Y. Oumi, K. Itabashi and T. Sano, *Micropor. Mesopor. Mater.*, 2006, **89**, 88.
14 D. Trong On and S. Kaliaguine, *Angew. Chem. Int. Ed.*, 2001, **40**, 3248.
15 R. Ryoo, C. H. Ko, M Kruk, V. Antochshuk and M. Jaroniec, *J. Phys. Chem. B.*, 2000, **104**, 11465.
16 M. Kruk, M. Jaroniec, C. H. Ko, R. Ryoo, *Chem. Mater.*, 2000, **12**, 1961.
17 M. Impéror-Clerc, P. Davidson and A. Davidson, *J. Am. Chem. Soc.*, 2000, **122**, 11925.
18 R. Ravishankar, C. Kirschhock, B. J. Schoeman, P. Vanoppen, P. J. Grobet, S. Storck, W. F. Maier, J. A. Martens, F. C. De Schryver and P. A. Jacobs, *J. Phys. Chem. B.*, 1998, **102**, 2633.
19 C. E. A. Kirschhock, R. Ravishankar, F. Verspeurt, P. J. Grobet, P. A. Jacobs and J. A. Martens, *J. Phys. Chem. B.*, 1999, **103**, 4965.
20 Y. Liu, W. Zhang and T. J. Pinnavaia, *Angew. Chem. Int. Ed.*, 2001, **40**, 1255.
21 Z. Zhang, Y. Han, L Zhu, R. W. Wang, Y. Yu, S. L. Qiu, D. Y. Zhao and F.-S. Xiao, *Angew. Chem. Int. Ed.*, 2001, **40**, 1258.
22 M. Llusar, G. Monrós, C. Roux, J. L. Pozzo, C. Sanchez, J. Mater. Chem., 2003, **13**, 2505.
23 X. G. Wang, K. S. K. Lin, J. C. C. Chan, S. F. Cheng, *J. Phys. Chem. B.*, 2005, **109**, 1763.
24 Y. M. Luo, Z. Y. Hou, R. T. Li and X. M. Zheng, *Can. J. Chem.*, 2007, **85**, 379.
25 K. Wang and X. Wang, *Micropor. Mesopor. Mater.*, 2008, **112**, 187.
26 J. C. Torres and D. Cardoso, *Micropor. Mesopor. Mater.*, 2008, **113**, 204.

PREPARATION OF ROD-SHAPE SILLIMANITE POWDER USING A JET MILL

G. Du, W. Cao and L. Liao

School of Materials Science and Technology, China University of Geosciences, Beijing, 100083, China
Email: dgx@cugb.edu.cn

1   INTRODUCTION

Light-weight refractory has many advantages, such as low heat conductivity, high refractoriness and acid resistance,[1-3] which can be wildly used as the lining of the high-temperature furnace and heating furnace. But it is well-known that its poor construction strength limits its application greatly. Fiber reinforcement is one of the feasible ways to relieve this problem.[4-6] The fibre used to enforce refractory should posses high refractoriness. Sillimanite powders are hence desirable candidate materials. A fibrous powder is usually prepared by both chemical and physical methods. The chemical method mainly includes chemical precipitation, sol-gel and microemulsion methods. Yin and Fujishiro[7] synthesized fibrous titania from hydrothermal reactions. Wang[8] produced fibrous nano-$SrTiO_3$ powders with a length/diameter (L/D) ratio of 20 with hydrothermal processes. High purity, small particle size and high L/D ratio are the characteristics of the powders prepared using the chemical methods. But they have poor yield with high costs. As an alternative, the physical method has a relatively high yield, and it is simple and inexpensive. For instance, Hashimoto[9] produced aramid fiber powders using a vibrating ball mill.

The preparation of rod-shape sillimanite powders has not been reported yet. In this paper, rod-shape sillimanite powders were prepared using a laboratory scale spiral jet mill and the processing parameters were optimized. In addition, the fragmentation mechanism of sillimanite crystals was discussed in detail.

2 EXPERIMENTAL

The raw material, sillimanite powder with a L/D ratio of 5.6 and d_{97}=80.81 µm, was provided by Linshou Mineral Ltd (Hebei Province, P. R. China). The spiral jet mill (QS-50 type) was manufactured by Shanghai Third Chemical Machine Factory.

The experimental procedure to produce roll-shape sillimanite powder is as follows. Firstly, the feed rate and grinding pressure of the jet mill is optimized. Secondly, the fragmentation mechanism is analyzed according to the crystal characteristics of sillimanite. When a sillimanite crystal is fragmented, the fissure must be developed in the face with the weakest bonding strength. Hence, in this paper, the least bonding strength was calculated from the crystal characteristics of sillimanite. The probability of the faces along which cracks occurred was then calculated.

SEM images were captured using a Quanta 600 FEI Electron Microscope with an acceleration voltage of 30 kV. The L/D ratio were measured from the SEM images of 500 randomly selected particles using a *Image plus 6.0 photo analysis software,* by defining the L/D ratio as the ratio of the total length to the total width. The XRD pattern was recorded with a PGENERAL model XD-2 X-ray powder diffractmeter (Cu Kα radiation, 40 kV, 30 mA and a scanning speed 2.0°(2θ)/min) . The particle size distribution was measured using a Laser Particle Size Analyzer (BT-9300H type).

3 RESULTS AND DISCUSSION

3.1 Analysis of the Raw Sillimanite Powder

The XRD pattern and SEM image of the raw material are shown in Fig 1. The main mineral phase in the raw material is sillimanite, and quartz is the main impurity. The SEM image shows that the maximum particle size is about 80 µm and most sillimanite particles have a rod shape with a L/D ratio of 5.4.

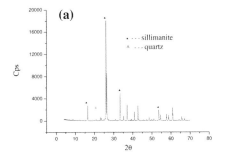

Figure 1   *XRD pattern (a) and SEM image (b) of the raw material*

3.2   Parameter Optimization

The raw sillimanite powder was smashed using the jet mill at different feed rates (0.13 g/s, 1.15 g/s, 2.53 g/s, 3.65 g/s, 4.20 g/s, 5.76 g/s) with the same feed pressure of 0.3 MPa and grinding pressure of 0.7 MPa. The effect of the feed rate on the L/D ratio and particle size was shown in Figure 2a.

As shown in Figure 2a, the particle size of the sillimanite powder increased gradually with the increase of feed rate. The L/D ratio of sillimanite particles increased significantly until the feed rate reached 3.65 g/s, and then, the L/D ratio ascended to 9.2. The L/D ratio kept constant when the feed rate changed from 3.65 g/s to 5.76 g/s. The volume of the chamber and the quantity of air flow was kept constant, and then the solid to gas ratio reached a suitable value when the feed rate reached 3.65 g/s. The increase of the feed rate did not change the solid/gas ratio in the chamber, but increased the discharge rate. The constant solid/gas ratio led to a constant L/D ratio of about 9. Due to the increase of the discharge rate, some powders were discharged without being sufficiently ground, so the particle size of the sillimanite powder increased. Therefore, the feed rate of 3.65 g/s was desirable in order to produce rod-shape sillimanite powders.

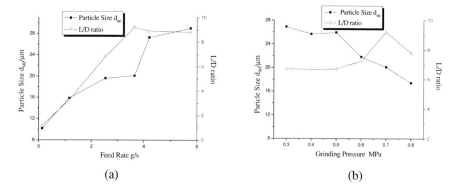

(a)   (b)

Figure 2   (a) *Effect of the feed rate on the particle size and L/D ratio of sillmanite powders;* (b) *Effect of grinding pressure on the d_{90} and L/D ratio of the sillimanite samples.*

The sillimanite powder was also smashed using the spiral jet mill at different grinding pressures (0.3 MPa, 0.4 MPa, 0.5 MPa, 0.6 MPa, 0.8 MPa,) with the same feed rate of 3.65 g/s and feed pressure of 0.3 MPa. The effect of the grinding pressure on the L/D ratio and particle size is shown in Figure 2b. As shown in Figure 2b, the particle size of the produced silllimanite powder decreased with the increasing grinding pressure. However, the L/D ratio reduced before it increased to a value of 9.2 with the increasing grinding pressure. The decreasing of particle size may be resulted from the higher grinding pressure, which caused the powder travelling at a higher velocity. The powder with a higher kinetic energy will be easier to be fragmented compared to that with a lower kinetic energy. So the optimized processing conditions to produce rod-shape sillimanite particles using the jet mill are at a feed rate of 3.65 g/s, with a grinding pressure of 0.7 MPa, and a feed pressure of 0.3 MPa.

3.3 Fragmentation Mechanism

The imposed kinetic energy must be greater than the chemical bonding strength when a sillimanite crystal is fragmented. In this paper, the crystal feature of sillimanite is analyzed and the minimal chemical bond energy in the perpendicular faces the three axes are calculated and the fragmentation mechanism is discussed.

The crystal structure of sillimanite plays an important role in the fragmentation process. The crystal structure is composed of chains of edge-sharing $[AlO_6]$ running parallel to the crystallographic c axis. These octahedral chains are cross-linked by double chains formed by alternate $[AlO_4]$ and $[SiO_4]$, which also run parallel to axis c (Figure 3a).[10,11] The sillimanite powder is rod-shape and the cleavage occurs on (010).

The chemical bond properties in the sillimanite crystal are listed in Table 1. As shown in Table 1, the Al-O bond and Si-O bond in the sillimanite crystal is primarily ionic bond. The coulomb forces between the cation and O^{2-} are an index to the strength of the bonds.

Table 1 *The characteristics of the bond in the sillimanite crystal.*

Bond Properties	Cation type		
	Al^{3+}	Al^{3+}	Si^{4+}
Accordinate Number	4	6	4
Ion Radius (Å)	0.53	0.62	0.40
Element Electronegativity	1.47	1.47	1.74
M^{n+}-O^{2-} Ionic Band Content (%)	64.3	64.3	53.9
M^{n+}-O^{2-} Bond Length(Å)	1.77	1.93	1.62
M^{n+}-O^{2-} Coulomb Force($\times ke^2$)	1.91	1.61	3.05
M^{n+}-O^{2-} Bond Angle	108°	90°	109°
M^{n+}-O^{2-} Bond Valence	0.71	0.46	1.02

There are two $[AlO_6]$ chains and two double chains formed by alternate $[AlO_4]$ and $[SiO_4]$ in the unit cell of the sillimanite crystal. If the fracture occurs on the *ab*-plane, the most possible fracture is along with plane *A* (Figure 3b). In the unit cell, it must break 2 Al-O bonds in $[AlO_4]$, 2 Si-O bonds and 2 $[AlO_6]$ chains at the edge-sharing place. So a total of $19.06ke^2$ is needed to fracture along with plane A. If the fracture occurs on *bc*-plane, the most possible fracture is along with plane *B* (Figure 3c). In the unit cell, it must break 2 Al-O bonds in $[AlO_4]$ and 2 Si-O bonds in $[SiO_4]$. A total of $9.92ke^2$ is hence required. If the fracture occurs on *ac*-plane, the possible fracture is along with plane

C (Figure 3d). In the unit-cell, it must break 4 Al-O bonds in [AlO$_6$]. Thus a total of 6.44ke^2 is needed. Since the least electrostatic force is need to break along with plane *C*, it is very likely that the sillimanite crystal will fracture along with the plane *C*. Consequently, the sillimanite crystal has the cleavage parallel to the c-axis and forms the rod-shape particles.

In the crystal structure of sillimanite, the Al-O bond connecting the double chains is the weakest and hence the easiest to break. The break of this bond creates a fracture face, which is perpendicular to the b-axis. It coincides with the macroscopic cleavage on (010) of sillimanite.

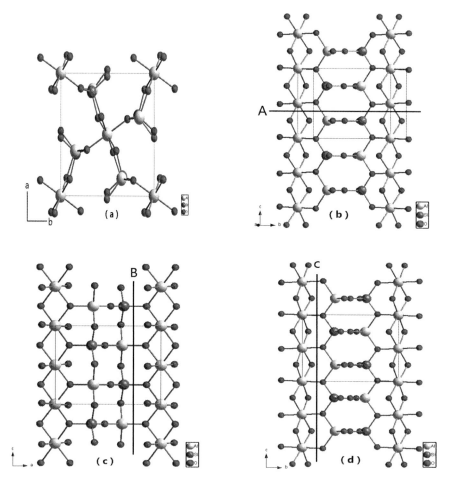

Figure 4   (a) *The typical crystal structure of sillimanite;* (b) *The most possible fracture of the sillimanite crystal on the plane perpendicular to c axis;* (c) *The most possible fracture of the sillimanite crystal on the plane perpendicular to b axis;* (d) *The most possible fracture of the sillimanite crystal on the plane perpendicular to a axis.*

If the imposed force applied to the unit cell of the sillimanite crystal is less than 6.44 ke^2, the sillimanite crystal will not break. If the imposed force is between 6.44 ke^2 and 9.92ke^2, the sillimanite crystal will break along with the plane C, and rod-shape particles will be obtained. If the imposed force is between 9.92 ke^2 and 19.02ke^2, the sillimanite crystal will break both along with the plane C and B, and the fraction of rod-shape particles will be reduced. If the imposed force is over 19.02ke^2, the sillimanite crystal will break along with the plane *A*, *B* and *C*, and the rod-shape particles can hardly be obtained.

The difference between the electrostatic force needed to fracture parallel to the c-axis and that perpendicular to the c-axis is not apparent. The ratio is about 1:3 (6.44 ke^2:19.02 ke^2). This may explain why it is difficult to get the sillimanite particles with high L/D ratios.

4 CONCLUSIONS

Rod-shape sillimanite powders were produced using a jet mill. With the optimized parameters (feed rate: 3.65 g/s, grinding pressure: 0.7MPa, feed pressure: 0.3 MPa), samples with a L/D ratio of 9.2 and d_{90}=18.7m were produced. The fragmentation mechanism was analysed on the basis of the sillimanite crystal feature. [AlO_6] connected double chains consisted of alternate [AlO_4] and [SiO_4]. The most possible fracture surface of the sillimanite crystal to prepare rod-shape particles is parallel to c-axis formed with the Al-O bonds in [AlO_6]. Since the ratio of the electrostatic force needed to fracture parallel to c-axis to that perpendicular to c-axis is relatively small (i.e., 1:3), it is difficult to produce sillimanite powders with high L/D ratios.

Acknowledgements

This project was funded by the Fundamental Research Funds for the Central Universities of China（No. 2010ZY48 and 2010ZY47) and the National Key Technology R&D Program of China (No.2011BAB03B06)

References

1   J. J. del Coz Diaz, P.J. Garcia Nieto, C. Beregon Biempica and M. B. Prendes Gero, *Appl. Therm. Eng.*, 2007, **8**, 1445.
2   O. Unal, T. Uyqunoglu and A. Yildiz, *Build. & Environ.*, 2007, **2**, 584.
3   R. Kozlowski, B. Mieleniak and P. Fiedorow, *Appl. Sci. & Manuf.*, 2005, **8**, 1047.
4   T. A. Parthasarathy, R. J. Kerans and S. Chellapilla, *Mater. Sci. & Eng.*, 2007, **1**, 120 .
5   G. Hilmas, A. Brady, U. Abdali and G. Zywicki, *Mater. Sci. & Eng.*, 1995,2, 63.
6   C. P. Ostertag, C. K. Yi and G. Vondran, *Cem. & Concr. Compo.*, 2001, **4** , 419.
7   S. Yin, Y. Fujishiro, J. Wu, M. Aki and T. Sato, *Mater. Process. Technol.*, 2003, **1**, 45.
8   J. S. Wang, S. Yin and T. Sato, *Mater. Sci. & Eng.*, 2006, **1**, 248.

9   A. Hashimoto, M.Satoh1, T.Iwasaki1 and M. Morita, *Mater. Sci.*, 2002, **37**, 4013

10 H. Yang and R. M. Hazen, *Phys. Chem. Miner.*, 1997, **325**, 9.

11 M. T. Vaughan and D. J. Weidne, *Phys. Chem. Miner.*, 1978, **3**,.133.

SURFACE MODIFICATION OF DIATOMITE WITH TITANATE AND ITS EFFECTS ON THE PROPERTIES OF REINFORCING NR/SBR BLENDS

J.H. Liao, G.X. Du, L.F. Mei, W.J. GUO , R.F. ZUO

School of Materials Science and Technology, China University of Geosciences (Beijing), Beijing, 100083, China. E-mail: dgx@cugb.edu.cn;   liaoshenglan2006@126.com

1   INTRODUCTION

Diatomite is a biogenic silicolites with porosity. Its main component is amorphous silica formed from the skeleton of single cell of diatom, and the chemical formula is $SiO_2 \cdot nH_2O$. In mineralogy, it belongs to opal-AG.[1,2] Owing to low cost, good chemical stability, safe and nontoxic, diatomite is often used as filter aid, adsorbent, catalyst carrier and functional filler.

Although precipitated amorphous silica is an important reinforcing agent for rubber, the problems such as high cost and severe pollution during its production process drive people to find the substitute to replace it. It is demonstrated that modified attapulgite and fly ash can substitute (or partly substitute) amorphous silica in rubber.[3] Diatomite, as a natural mineral, is rich in resources and of low cost. Diatomite and precipitated amorphous silica share the same chemical composition, and they both posses porous structure and appear acidic on the surface.[4] In addition, diatomite added into rubber or plastic can enhance its rigidity, hardness and wear-resistance significantly.[5] As a filler, it can obviously increase strength, aging, curability, recovery, compression, permanent deformation and workability of fluorubber.[6]   Therefore, substituting amorphous silica with diatomite to reinforce rubber is significant. Due to its low dispersity in organic matrix and poor compatibility in polymer matrix caused by high surface energy and hydrophilicity, diatomite should be surface modified first. Diatomite modified by stearic acid was used in natural rubber/styrene butadiene rubber blends (NR/SBR). The results showed that diatomite took on good dispersion in organic matrix and enhanced mechanical properties of rubber.[7] At present, the study of surface modification of diatomite is aiming at improving the adsorb-ability in wastewater treatment,[8~10] while little in reinforcing rubber.

In this paper, diatomite was surface modified using titanate. The dosage of titanate, mechanical properties of NR/SBR blends with modified diatomite were examined and the underlying mechanism was explored as well.

2   EXPERIMENTAL

2.1   Materials

Raw diatomite sample (d_{50}=15.64 μm，d_{90}=24.43 μm) was obtained from Changbai Mountain of Jilin Province, China. The titanate coupling agent (YB-502) was supplied by Changzhou Yabang Yayu Auxiliaries Co., Ltd. Titanate was added to diatomite with constant stirring for 90 min at 90 ℃ with a stirring speed of 1500 RPM. The modified diatomite samples were evaluated in terms of the oil absorption value and the sedimentation time in kerosene. The modified diatomite was extracted in isopropanol for 9 h to remove the titanate which does not react. The modified diatomite was mixed into NR/SBR blends as a reinforcing agent. The properties of rubber were detected after vulcanization 10 min at 143±1 ℃.

The curing formula (mass ratio) are as follow: natural rubber 80 g, styrene-butadiene rubber 20 g, sulfur 2.2 g, accelerator M 0.8 g, accelerator DM 1.0 g, accelerator CZ 0.8 g, ZnO 5 g, stearic acid 0.7 g, antioxidant WL 1.0 g, antioxidant PF 2 g , antioxidant MB 0.8 g, light calcium carbonate 35 g, polyethylene glycol 1 g, high efficient dispersant AT-C 2 g, $AlSiO_3$，20g；resin RX-80 1.5 g and reinforcing agent 20g.

2.2   Characterization

The particle size distribution was analyzed using laser particle size analyzer (BT-9300 type), the property of reinforced rubber was tested using a tensile testing machine (XLL-250), The surface properties of diatomite samples were analyzed using Fourier transform infrared spectrometer (PE983) and X-ray photoelectron spectrometer (PHI Quantera SXM), and the surface morphology of samples were examined with scanning electron microscopy (QUANTA600).

3   RESULTS AND DISCUSSIONS

3.1   Evaluation and Characterization of Modified Diatomite

3.1.1 Oil absorption value and settling time. The influence of titanate dosage on diatomite surface modification effect was evaluated in terms of oil absorption value and sedimentation time in organic solution, as shown in Figure 1. When 1.2% titanate was used, the sedimentation time increased from 1 min to 6 min, and the oil absorption decreased to the minimum (1.34 ml/g).

3.1.2 Characterization of modified diatomite and mechanism of action. The FTIR spectra of samples were shown in Figure 2. As shown in Figure 2, the diatomite is an amorphous [SiO$_4$] tetrahedron structure and there is small content of adsorbed water in the holes of diatomite.[8]   New peaks of C-H bond (at 2961 cm^{-1}, 2932 cm^{-1}and 2875 cm^{-1}) in the diatomite modified with titanate appeared, but they did not disappear in diatomite extracted by isopropanol. The intensity of the 1630 cm^{-1} decreased and shifted obviously, indictaing that the chemical reaction between diatomite and titanate occurred.

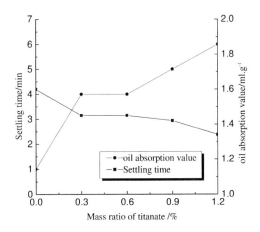

Figure 1   *Effect of titanate dosage on the surface treatment of diatomite*

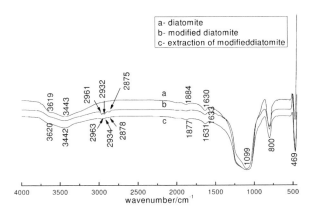

Figure 2   *FTIR spectra of different diatomite samples*

The X-ray photoelectron spectrometer (XPS) spectra in the Si(2p) and O(1s) regions were presented in Figures 3 and 4. The spectra obtained are deconvoluted into multiple sub-peaks of Si(2p) and O(1s) contained in different functional groups using Gaussian-Lorentzian fit. The peaks of Si(2p) represent the Si in Si-O-Si (103.1 eV) and Si-OH (104.2 eV)，respectively, which occurred due to chemical shifts through modification. The rest peak (103.9 eV) of modified sample was attributed to Si-O-Ti, meaning that the chemical environment of Si was changed. The binding energy of O(1s) in Figure 4 corresponds to the possible functional groups as follows: 532.3 eV vs.Si-O-Si, 533.4 eV vs. Si-OH (Figure 4a). After modification, the binding energy of O(1s) in Si-O-Si decreased and the new peak corresponds to the O of Si-O-Ti (532.7eV). Because Ti atom with lower electro-negativity replaced H atom, the electron density of O and shielding effect increased and thus the binding energy of O(1s) decreased.

Diatomite is a biogenic silicolites containing a mass of surface hydroxyl group, The general formula of titanate coupling agent is $(RO)_M$-Ti-$(OX$-R'-Y$)_N$ $(1 \leq M \leq 4, M+N \leq 6)$. The organic monolayer can be formed on the surface of diatomite due to the reaction of the part of $(RO)_M$ and hydroxyl,[9] the mechanism of action is shown in Figure 5.[10,11]

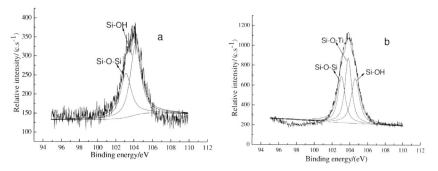

Figure 3   *XPS Si(2p) element fitting spectra of diatomite (a) before and (b)after modification*

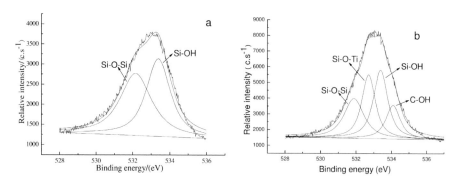

Figure 4   *XPS O(1s) element fitting spectra of diatomite (a) before and (b) after modification*

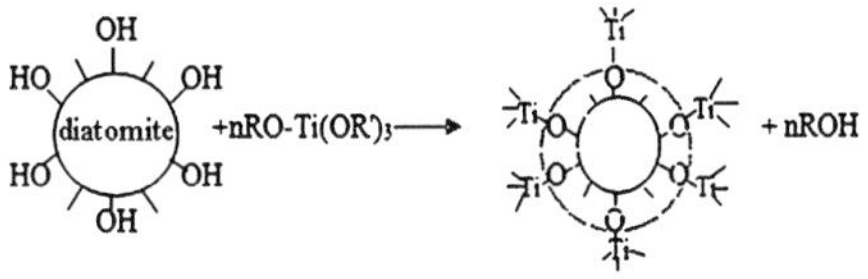

Figure 5 *The modification mechanism of titanate on the surface of diatomite particles*

3.2 Effect of Modified Diatomite on Properties of NR/SBR Blends

The mechanical properties of NR/SBR blends filled with different reinforcements were shown in Table 1. To the best of our knowledge, the oxyalkyl of titanate could combine with hydroxyl of diatomite easily, meanwhile, the organic molecular chains of titanate and carbon chains of rubber had chemical reactions or physical entanglement, so the tear strength and stress at 300% elongation increased. The reason that the elongation for NR/SBR filled with modified diatomite decreased is because a high degree of cross-linking to the disadvantage of the orientation along tensile direction is induced. [12]

Table 1 *Mechanical properties of nature rubber/ styrene butadiene rubber (NR/SBR) blend with different reinforcements*

Material	Tensile Strength /MPa	Elongation at break /%	Permanent set /%	Tear strength /(KN/m)	stress at 300% elongation /%	shore A hardness
Diatomite (non-modified)	14.9	576	28	32.5	5.0	63
Diatomite (modified by titanate)	13.2	427	28	36.6	7.2	68
Precipitated amorphous silica	21.1	720	40	43.4	4.4	68

In Figure 6, SEM images of cross-sections of diatomite filled NR/SBR samples are shown. Figures 6d and 6e show the SEM images of fresh section of rubber vulcanized, only adding diatomite, vulcanizer and accelerant, which was brittle and fractured in liquid nitrogen. The polymer molecular changed into smaller weight molecular due to the mechanical shear force and was squashed in macropore of diatomite during mixing. Then the rubber molecular cross linking are presenting reticulation structure after vulcanization. The fine diatomite particles are non-uniform in the rubber matrix and absorb other auxiliaries (Figure 6b). After modified diatomite was filled in NR/SBR, the vulcanized rubber is smoother. Diatomite disperses well in rubber (Figures 6c, 6e), and the interface with polymer matrix is dim. The results demonstrate that the modified diatomite have good associativity with the rubber matrix.

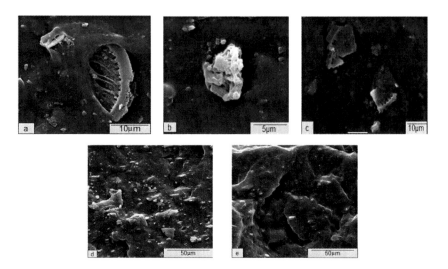

Figure 6   *SEM photos of NR/SBR blends with diatomite (a, b, d) and modified diatomite (c, e)*

4   CONCLUSIONS

Diatomite modified with titanate can improve its dispersion in organic matrix, and the optimal dosage of titanate was 1.2%. The adsorption type between titanate and diatomite is chemical adsorption and the Si-O-Ti was formed. Diatomite can reinforce the rubber to some extent. When modified diatomite was added to the NR/SBR blends, the tearing strength and 300% elongation of NR/SBR increased by 12.6% and 44.0%, respectively, while the tensile strength and the elongation at break decreased by 25.9% and 11.4%, respectively. The interaction among particle was weakened due to the titanate molecule and the organic molecule of diatomite surface modified with titanate. Therefore, the interaction between diatomite and rubber was strengthened, the cross-linking density of rubber increased, and the ability of elastic deformation decreased.

Acknowledgements

Financial support from Fundamental Research Funds for the Central Universities (No. 2010ZY47,No.2010ZY48, No.2011PY0167) is gratefully acknowledged.

References

1   P. Yuan, D.Q. Wu, Z.Y. Lin, et al. *Applied Surface Science*, 2004, **227**, 30.
2   W.S. Xiao, W.S. Peng, G.X. Wang, et al. *Spectrosco Spect Anal*, 2004, **24**, 690.
3   S.Z.Peng, L.Song, B.C.Ma, et al. *Journal of the Chinese ceramic society*, 2009, **37**,103.
4   P. Yuan. *Guangzhou: GuangZhou Institute of Geochemistry, Chinese Academy of Sciences*, 2001.
5   D.Q. Zhang, W.W. Lin, F. Chen, et al. *Materials Science &Engineering*, 1997, **15**,61.
6   B.K. Wang, Y. Zhang. *Special Purpose Rubber Products (in Chinese)*, 2010, **31**, 26.
7   J.H. Liao, G.X. Du, Q. Xue, et al. *J. of the Chinese Ceramic Society*, 2011, **39**, 641.
8   P. Yuan, D.Q. Wu , Z.Y. Lin, et al. *Spectrosc Spect Anal (in Chinese)*, 2001, **21**,784.
9   S.L. Zheng. *Beijing: China Building Material Industry Publishing House*, 2004, 51.
10 M.K. Wei , X.F. Wang, J.M. Song, et al. *New Chemical Materials*, 2003, **31**,40.
11 Y.L. Lin, T.J. Wang, C. Qin, et al. *Acta Phys-Chim Sin (in Chinese)*, 2001, **17**,170.
12 Q.Z. Yang. *Beijing: Chemical Industry Press, (in Chinese)*, 2005, 92.

Characterisation

ESTIMATION OF 3D FACETED GROWTH RATES OF POTASH ALUM CRYSTALS IN A HOT-STAGE REACTOR USING ONLINE 2D IMAGES

C. Y. Ma, J. Wan and X. Z. Wang

Institute of Particle Science and Engineering, School of Process, Environmental and Materials Engineering, University of Leeds, Leeds LS2 9JT, UK

1 INTRODUCTION

Particle shape is extremely important for many industrial applications including pharmaceuticals and fine chemicals. In recent years, on-line high speed imaging[1-12] has proved to be a very promising technique for real-time measurement of particle shape. However, current imaging technique is limited to providing two-dimensional (2D) information of particle shape. On the other hand, a morphological population balance (PB) model[13] has been developed to predict the evolution of individual crystal faces, hence the shape evolution of crystals in a reactor. However, the faceted growth rates as the input for the morphological PB model are decisive and also difficult to obtain using current imaging techniques. Therefore, it is important to accurately estimate faceted crystal growth rates in order to predict crystal shape evolution, hence crystal size distributions and other properties, for optimisation and control of crystallisation processes using the morphological PB method.

Experimental studies[14, 15] demonstrated that the morphology of potash alum is dominated by 8 {111}, 6 {100} and 12 {110} faces (Figure 1).[13] Through single crystal studies, the relative growth rates between the {111}, {100}, {110} faces were found to be roughly 1:5.3:4.8.[14] X-ray topographic studies[15] were also used to estimate the faceted growth rates at different relative supersaturation levels. The growth rate of the {111} face and also the overall growth rate of a single potash alum crystal at a constant temperature were measured and correlated as a function of the relative supersaturation.[16] However, there were no correlations available for {100} and {110} faces. In a recent study,[13] the faceted growth rates were estimated based on the experimentally-obtained correlations for face {111} and their corresponding rate ratios against face {111} for faces {100} and {110} due to the lack of reliable experimental data. Therefore it has been realised that the accurate estimation of crystal growth rates of individual faces is essential if the control of crystal shape and size distributions are to be achieved.

In this paper, the faceted growth rates of potash alum crystals were studied using a hot-stage microscopic system.[1, 12, 17] A multi-scale segmentation algorithm[18] and a line detection method[19] were used to process the recorded 2D images with the extracted 2D information for deducing the normal distances of {111}, {100} and {110} faces, as shown in Figure 1. The obtained growth rates were compared with experimental data from literature.

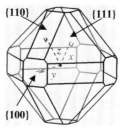

Figure 1   *Morphology and three normal distances (x, y, z) of a potash alum crystal.*[13]

2 APPARATUS AND EXPERIMENTAL TECHNIQUES

An imaging system[1, 12] was used to record on-line 2D images of potash alum crystals grown in a Linkam hot-stage[17] (Figure 2). The camera was placed vertically over the hot-stage reactor and the strobe light below the reactor. Details of the on-line imaging and hot-stage systems can be found in literature.[1, 12, 17] Saturated solution at 40°C was prepared with 23.7 g of potash alum in 100 ml of distilled water. The hot-stage was maintained at 40°C for half hour with an empty sample holder. Then 0.1 mL of the saturated solution was injected into the holder and cooled down from 40 to 25°C at a cooling rate of 3°C/min. The captured images were stored for further analysis.

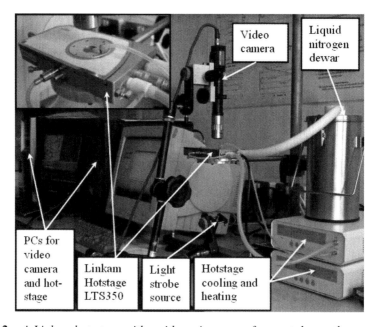

Figure 2   *A Linkam hot-stage with a video microscopy for crystal growth measurement.*

3 ESTIMATION OF FACETED GROWTH RATE

3.1 Image Analysis

The recorded images of crystals were segmented at each sampling instant using a multi-scale segmentation algorithm.[18] Line detection was then performed to obtain the polygonal representation. The distances between various pairs of parallel lines were calculated based on the polygonal representation. These distances are closely related to parallel plane distances of the corresponding 3D crystal. Figure 3 shows the sequence of segmented 2D images with line detection for growth rate measurement.

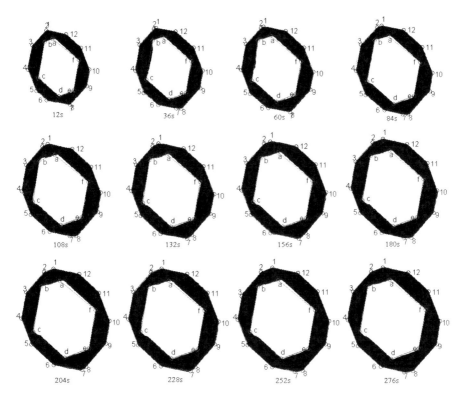

Figure 3 *Segmented 2D images of potash alum crystals with line detection for growth rate measurement*

3.2 Estimation of Faceted Growth Rates

The potash alum crystals have positioned themselves with the lowest free energy in such a way that the largest {111} face is perpendicular to the vertical axis. The area of the parallel surfaces of face {111} was recorded on an image as white colour with the other areas as black colour due to the diffraction of the light passing through the angled surfaces (Figure 4(a)). From this viewpoint, the 2D images obtained have twelve outer edges and six inner edges.

By perpendicularly cutting the potash alum crystal (Figure 1) through 4 {111}, 2 {110} and 2 {100} faces, the obtained face is shown in Figure 4b. The distance D_1 in Figure 4a, is the projected length of line $\overline{ae}$ on {111} face with a projecting angle, $\beta+\alpha$ (β is obtained from morphology calculation and $\alpha = tg^{-1} \{ctg(\beta)-D_1/[D_2 \sin(\beta)]\}$). D_2 is the projected length of line $\overline{ae}$ with an angle, α. D_3 is the summary of D_1 and the projected lengths of lines $\overline{cd}$, $\overline{de}$, $\overline{gh}$ and $\overline{ah}$ on the {111} face with the corresponding angles of 2β-90°, β, 2β-90° and β, respectively. D_1, D_2, D_3 are calculated from the coordinates of the corresponding points as shown in Figure 4a. With mathematical manipulations, the three normal distances (x, y and z) as shown in Figure 1 are estimated:

$$x = D_1/2 \; tg(\beta+\alpha) \tag{1}$$

$$y = \frac{D_3 - D_1 - 2D_2 tg(\alpha)\sin(\beta)}{2\sin(2\beta-90^o)}\sin(\beta) + \frac{D_2}{2}tg(\alpha) \tag{2}$$

$$z = D_2/2 \tag{3}$$

The growth rates of {111}, {100} and {110} faces, can be obtained using the estimated values of x, y and z. Noting that the method developed here to obtain the growth rates of individual faces can be used for potash alum crystals and also other compounds with similar shape. This study provides a feasible method for the first time to directly estimate the growth rates of {111} face, and also {100} and {110} faces.

In the hot-stage experiments, a spherical cap etched on a glass slide was used to grow potash alum crystals using a 0.1 mL of potash alum solution. The solid weight of crystals was estimated using the obtained three normal distances. The relative supersaturation, σ, was then estimated with solid weight, solution volume and solubility.[13]

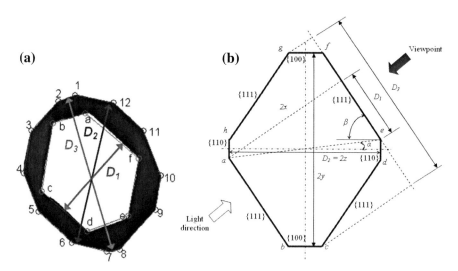

Figure 4   *(a) crystal image from experiment. (b) new surface via perpendicularly cutting a potash alum crystal (Figure 1) through 4 {111}, 2 {110} and 2 {100} faces.*

4   RESULTS AND DISCUSSION

Figure 5(a) shows the distributions of normal distances of {111}, {100} and {110} faces calculated using Eqs (1)-(3). The normal distances increased with time and a second order polynomial function was used to curve-fit them with R^2 being over 0.99.

By using the estimated relative supersaturation and the curve-fitted normal distance distributions in Figure 5a, the crystal growth rates of three individual faces against the relative supersaturation can be obtained (Figure 5b) as follows:

$$v_{\{111\}} = 19.27 \times 10^{-6} \, \sigma^{\, 2.24} \tag{4}$$

$$v_{\{110\}} = 35.24 \times 10^{-6} \, \sigma^{\, 2.36} \tag{5}$$

$$v_{\{100\}} = 6.86 \times 10^{-6} \, \sigma^{\, 1.39} \tag{6}$$

Figure 6 shows the 3D faceted growth rates of potash alum crystals obtained from the current study and literature. The growth rate of {111} face estimated in this study is close to the measured results in literature.[16, 20] However, there are no reliable experimental data in literature for {111} and {100} faces. The relative growth rates of {111}, {100} and {110} faces are 1:5.3:4.8 at 30°C [14] and the relative growth rates of faces {111} and {100} at 32°C [16] are 1:2.16 for small crystals and 1:1.16 for large crystals. From this study, the relative growth rates of the three faces are 1:1.62:1.46 at 30°C and 1:1.55:1.47 at 32°C, which are close to the ratios in literature. Noting that the obtained growth rates of {111}, {100} and {110} faces provide an important input to the morphological PB model[13] for crystallisation process simulations.

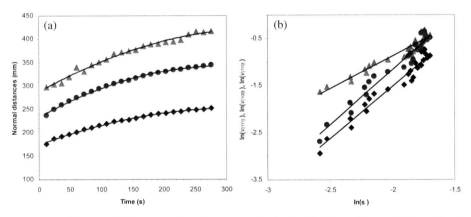

Figure 5   *(a) normal distances from image analysis (symbols) and curve-fitting (lines); (b) faceted growth rates (triangles — {100} face; circles — {110} face; diamonds —{111} face)*

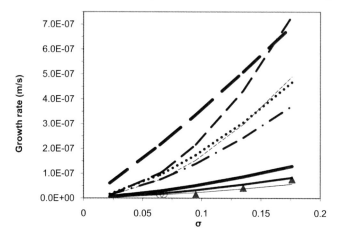

Figure 6   *Growth rates of potash alum crystals (Dash lines: thin thickness — $v_{\{111\}}$, normal thickness — $v_{\{110\}}$, thick thickness — $v_{\{100\}}$; Solid lines[13]: thick thickness — {111} face, normal thickness — {100} face, thin thickness — {110} face; ○ — {111} face[15]; ▲ — {111} face[21]; Dot line — {111} face[20]; Dash and dot line —{111} face[16]).*

5   CONCLUSION

A morphological PB model can predict crystal shape and size distributions for real-time optimisation and control of crystallisation processes. However, the most important input for this model is the faceted growth rates, which are extremely difficult to be measured using current PAT techniques. In this paper, a 2D imaging system was used to obtain 2D online images of potash alum crystals in a hot-stage reactor and mathematical formulation was developed to extract normal distances of individual faces from the 2D images. The obtained 3D faceted growth rates were compared with the available experimental data with promising results.

Acknowledgements

Financial support from the UK Engineering and Physical Sciences Research Council (EPSRC) for the projects of Shape (EP/C009541) and StereoVision (EP/E045707) are acknowledged. We also thank the industrial collaborators: AstraZeneca (Dr G Steele, Dr M J Quayle), Malvern Instruments (Mr F McNeil-Watson, Dr J Corbett, Mr D Watson, Dr D Roberts), National Nuclear Laboratory (Dr D Rhodes), Pfizer (Dr I Marziano, Prof R Docherty), Syngenta (Dr N George) and 3M HealthCare (Dr C Blatchford).

References

1. J. Calderon de Anda, X. Z. Wang, X. Lai, K. J. Roberts, K. H. Jennings, M. J. Wilkinson, D. Watson and D. Roberts, *AIChE J.*, 2005, **51**, 1406-1414.

2. E. Kougoulos, A. G. Jones, K. H. Jennings and M. W. Wood-Kaczmar, *J. Cryst. Growth*, 2005, **273**, 529-534.
3. C. Y. Ma, X. Z. Wang and K. J. Roberts, *Adv. Powder Technol.*, 2007, **18**, 707-723.
4. M. Oullion, F. Puel, G. Fevotte, S. Righini and P. Carvin, *Chem. Eng. Sci.*, 2007, **62**, 820-832.
5. H. Y. Qu, M. Louhi-Kultanen and J. Kallas, *J. Cryst. Growth*, 2006, **289**, 286-294.
6. X. Z. Wang, J. Calderon De Anda and K. J. Roberts, *Chem. Eng. Res. Des.*, 2007, **85A**, 921-927.
7. X. Z. Wang, J. Calderon De Anda, K. J. Roberts, R. F. Li, G. B. Thomson and G. White, *KONA*, 2005, **23**, 69-85.
8. Y. Zhou, X. T. Doan and R. Srinivasan, Joint 16th ESCAPE and 9th PSE: Computer - Aided Chemical Engineering 21A, Garmisch-Partenkirchen, Germany, 2006.
9. J. Scholl, D. Bonalumi, L. Vicum, M. Mazzotti and M. Muller, *Cryst. Growth Des.*, 2006, **6**, 881-891.
10. X. Z. Wang, K. J. Roberts and C. Y. Ma, *Chem. Eng. Sci.*, 2008, **63**, 1171-1184.
11. L. X. Yu, R. A. Lionberger, A. S. Raw, R. D'Costa, H. Wu and A. S. Hussain, *Adv. Drug Deliver. Rev.*, 2004, **56**, 349-369.
12. M. J. Wilkinson, Jennings, K. H., Hardy, M., *Microsc. Microanal.*, 2000, **6**, 996-997.
13. C. Y. Ma, X. Z. Wang and K. J. Roberts, *AIChE J.*, 2008, **54**, 209-222.
14. H. Klapper, R. A. Becker, D. Schmiemann and A. Faber, *Cryst. Res. Technol.*, 2002, **37**, 747-757.
15. R. I. Ristic, B. Shekunov and J. N. Sherwood, *J. Cryst. Growth*, 1996, **160**, 330-336.
16. J. W. Mullin and J. Garside, *T. I. Chem. Eng.-Lond.*, 1967, **45**, T285-T290.
17. Linkam Scientific Instruments, *http://www.linkam.co.uk/*, 2010.
18. J. Calderon de Anda, X. Z. Wang and K. J. Roberts, *Chem. Eng. Sci.*, 2005, **60**, 1053-1065.
19. J. Wan, C. Y. Ma and X. Z. Wang, *Particuology*, 2008, **6**, 9-15.
20. M. Matsuoka, Y. Abe, H. Uchida and H. Takiyama, *Chem. Eng. Sci.*, 2001, **56**, 2325-2334.
21. S. Nollet, C. Hilgers and J. L. Urai, *Geofluids*, 2006, **6**, 185-200.

A NOVEL APPROACH TO QUANTIFYING AND SCALING AIR CURRENT SEGREGATION FROM EXPERIMENTS TO INDUSTRIAL SILOS

S. Zigan

Chemical & Process Engineering, University of Surrey, Guildford, GU2 7XH, UK

1 INTRODUCTION

Air current segregation (ACS) is a segregation mechanism which separates fines (defined as particles smaller than 42 microns) from the main bulk during a silo filling procedure. Fines are carried away from the particle jet by circulating fluids in a silo. The high concentration of fines accumulating near a silo wall can cause severe problems in the downstream processes.[1] The amount of fines accumulating near the silo wall relates to the spread of the particle jet. This in turn is influenced by the dynamic interaction with the surrounding fluid in the silo and depends on parameters such as the free fall height of the powder and the diameter of the particle jet.[2]

These parameters have to be grouped into dimensionless numbers applying dimensional analysis (DA).[3] Dimensional analysis provides a scaling rule which is satisfying the requirement of dimensional and dynamic similarity between different sized silos. Applying the scaling rule to an industrial silo filled with alumina provides the scale down parameters for an experimental silo. The experimental silo is a water silo filled with sand particles. When the dynamic in the two silos is similar the ACS should be similar as well.

2 LITERATURE REVIEW

The scaling rule between the industrial alumina silo and the experimental water silo has to incorporate the fluid-phase interaction with the falling particle jet. The falling particle stream in the industrial silo spreads because air becomes entrained in the particle jet and causes a diffusion of the jet.[2] Ansart et al.[4] found that the voidage between the particles in the jet is larger with increasing drop height of the powder. The same applies for the radius of the dust laden boundary layer which forms when fines start to create a miscible plume around the particle jet. For each particle jet a characteristic radius could be obtained.[5] Ansart et al.[5] suggested using this radius at various silo heights to estimate a spread angle. The spread angle is used in this paper to compare the dynamic of the particle air-flow in different sized silos.

Applying DA requires that the dynamic of the free falling particles in the industrial silo should be the same as that in the water silo. One way of generating a similar particle jet behaviour in the two different silos was to scale the sand particle diameter according to the terminal velocities of the alumina particles. This created a particle velocity profile which was similar in both the industrial silo and the water silo. The velocities of the settling particles were close to the single particle terminal velocity because they settled as a plume with a large voidage between the particles, as the drop height was significant in the industrial silo.[2]

The single particle terminal velocity included parameters such as fluid density and viscosity which would determine the size of the sand particles in the water experiment. Other parameters affecting the dynamic of the particle jet were the powder feeding rate and the fluid extraction rate from the silo.[6]

3 METHODOLOGY

The scaling experiments were carried out in an industrial steel silo with a diameter of 10 metres and a height of 30 metres. Alumina was fed via air slides at a feeding rate (S) of around 1000 kg/min. The powder was fed centrally into the silo and the air was extracted from the silo at an air extraction rate (E) of around 20,000 l/min. The particle jet in the industrial silo was video recorded and samples were taken with a sample stick near the silo wall and 1 metre away from the silo wall. Four samples of 200 g were collected from around 20 cm below the heap surface in the 80 per cent filled industrial silo.

The cylindrical water silo was scaled down from the industrial silo by a scaling factor of 70. The scaling law is based on dimensional analysis and is presented equation 1:

$$I_s = f\left(\frac{E}{S}, \frac{R^2 v_t}{S}\right) \tag{1}$$

where I_s is the segregation index defined in[6], E is the air extraction rate, S is the powder feeding rate, R the representative geometry dimension (*e.g.* the radius of the silo) and v_t the terminal velocity of a single particle. Parameters such as the particle and fluid densities were included in the single particle terminal velocity, which provided a single value for scaling the dynamic of the particle jet between different sized silos. This simplifies the DA. A single particle terminal velocity was chosen because the particles settled in the water silo with a large enough voidage between them.

Table 1: *Values for scaling parameters between the industrial silo and the water silo (using a scaling factor of 70)*

	Industrial silo	Water silo
Air extraction rate (E) in l/min	20000	4
Powder feeding rate (S) in kg/min	1000	0.2
Radius of silo (m)	10	0.15
Terminal velocity (v_t) microns	0.14	0.14

The alumina particle diameter used for calculating the terminal velocity in the industrial silo was 42 microns. The density of alumina particles was 1500 kg/m³ according to the data sheet of the supplier. The corresponding sand particle diameter was calculated at 500 microns matching the single particle terminal velocity of the alumina particles (42 microns). The powder fed into the industrial silo contained around 7 mass per cent of fines and the water silo around 8 mass per cent. The sand particles were scaled according to the alumina particle size distribution in the industrial silo. The scaled parameters for the water silo are presented in Table 1:

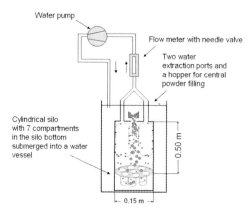

Figure 1 *A diagram showing how the water silo was fully submerged in another vessel*

    The experimental set up of the water silo included a cylindrical section made of Plexiglas which had a diameter of 15 cm and a wall thickness of 2 mm. The whole silo was submerged in a 500 litre plastic vessel, which was filled completely with water (see Figure 1). A centrifugal pump pumped the water from the water silo back into the vessel. Plastic pipes connected the experimental silo with a flow meter and a needle valve to control the fluid flow extracted from the water silo. The flow meter had a range of 20 to 270 l/h[1] $\pm$ 1.2 per cent. A closed system was created which kept the water level in the vessel constant. The water level was above the material feeding hopper where sand was submerged in the water before being fed into the silo. In the vessel, a side glass was installed to observe and video record the particle jet in the cylindrical silo. The pictures were analysed and the spread angle was estimated and compared with the spread angle of the particle jet in the industrial silo. The spread angle α was calculated using the following equation adopted by Ansart *et al.*:[5]

$$\alpha = 100 \times (r_0 - D_0)/h \qquad (2)$$

where α is the spread angle, r_0 is the radius of the jet, D_0 is the outlet diameter of the hopper and h is the drop height.

    Samples collected from the compartments of the water silo were analysed and, using a simple statistical test,[7] compared with the data from the industrial silo (see Equations 3, 4 and 5):

$$s^2 = \frac{1}{n-1}\sum_{i=1}^{n}(x_i - \bar{x})^2 \qquad (3)$$

---

[1] 1 l/h around 2.7×10^{-7} m^3/s

Where

$$\bar{x} = \frac{1}{n}\sum_{i=1}^{n} x_i \qquad (4)$$

and

$$\varepsilon = \frac{s}{\bar{x}} \qquad (5)$$

where s is the standard deviation for a series of n samples and x_i is the property of the sample i *e.g.* the mass fraction of fines in the sample. By dividing the standard deviation s by the mean value $\bar{x}$, the coefficient of variation ε was obtained.

4   RESULTS AND DISCUSSION

The data obtained from the experiments were analysed by comparing the coefficient of variation (ε) between the water silo and the industrial silo (see Table 2). The values of the coefficient of variation were found to be similar between the industrial silo 0.37 and the water silo 0.3. This confirmed the video analysis of the two particle jets in the two silos. When the nearly empty alumina silo was filled and the particle jet video recorded, it was found after 1 min. that the air was laden with fines and that the visibility in the silo was zero (a 'snowstorm effect'). The particles in the water silo behaved in a similar way where fines started to circulate in the fluid immediately after the feeding procedure started. The particle dynamics in the different sized silos were therefore found to be similar.

   The spread angle provided another method for analysing whether there was any similarity with respect to the particle jets and the particle dynamic in the water silo and the industrial silo. Applying equation 2 (above), it was found that the estimated spread angle (α) was 16 % for the industrial silo and 14 % for the water silo (see Table 2).

   Spread angles of around 6 per cent were reported by Ansart *et al.*[4] which implies that the dynamic of the particle jet in their equipment was different to the one in our water and industrial silos. This shows that valuable information about the particle jet can be correlated between the water and industrial silo.

   The statistical results and also the spread angle calculations seem to support the soundness of the scaling rule.

Table 2: *The particle jets in the water silo and the industrial silo were compared by estimating the coefficient of variation and the spread angle.*

	Industrial silo	Water silo
Standard deviation (s)	16.26	8.99
Mean value ($\bar{x}$)	43.5	30.3
Coefficient of variation (ε)	0.37	0.30
Radius of particle jet r_0 in m	1.3	0.02
Drop height in m	7	0.1
Orifice diameter D_0 in m	0.163	0.006
Spread angle (α) as a %	16	14

a) Water silo   b) Industrial alumina silo

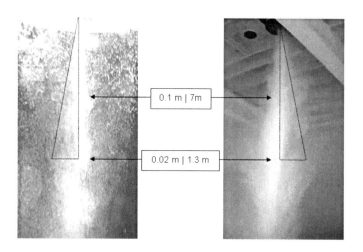

Figure 2   *Comparing the particle jets and giving values for calculating the spread factor for (a) the water silo and (b) the industrial alumina silo (applying a scaling factor of 70)*

5 CONCLUSIONS

It can be concluded that the particle jet in the water silo and the industrial silo are similar when the proposed scaling rule based on DA is applied. The particles settled as a plume in both the water and industrial silos. The video analysis of the two particle jets showed a similar particle flow in both silos. The similar results for the coefficient of variation indicated that the separation of fines from the main bulk occurred in the two silos at the same magnitude. The spread angle as a qualitative indicator was within the same range for both silos which suggests that the three dimensionless groups can be applied successfully to scale experimental findings to industrial silos.

The findings are giving confidence in the scaling rule. They support the DA approach to predict ACS in small scale equipment and correlate it later to industrial applications.

Future research could aim to develop a correlation model which can predict ACS. This would be of great value for practitioners because ACS for certain types of powders could be directly predicted from charts.

Acknowledgements

The author wishes to thank members of Tel-Tek (POSTEC), Norway, for their technical and financial support for this project and also Professors R.B. Thorpe, U. Tuzun and G. Enstad for their advice and encouragement.

References

1   J. Johanson, *Bulk Sol. Handl.*, 1987, **7**, 205.
2   O. Koichiro, F. Katsuya and T. Yuji, *Powder Techn.*, 2001, **115**, 90.

3 S. Zigan, R.B. Thorpe, U. Tuzun and G. Enstad, *Part. & Part. Sys Char.*, 2007, **24**, 124.
4 R. Ansart, A. Ryck, J. Dodds, M. Roudet, D. Fabre and F. Charru, *Powder Techn.*, 2009, **190**, 274.
5 R. Ansart, A. Ryck, J. Dodds, *Chem. Eng. Journ.*, 2009, **152**, 415.
6 S. Zigan, R.B. Thorpe, U. Tuzun, G. Enstad and F. Battistin, *Powder Techn.*, 2008, **183**, 133.
7 M. Stroeven, H. Askes and L.J. Sluys, *Comput. Methods Appl. Mech. Engrg.*, 2004, **193**, 3221.

INVESTIGATION INTO THE DEGREE OF VARIABILITY IN THE SOLID STATE PROPERTIES OF COMMON PHARMACEUTICAL EXCIPIENTS - MICROCRYSTALLINE CELLULOSE

J.F. Gamble[1], W.-S. Chiu[1, 2], V. Gray[1], H. Toale[1], M. Tobyn[1] and Y. Wu[3]

[1] Exploratory Biopharmaceutics R&D, Bristol-Myers Squibb, Reeds Lane, Moreton, Wirral, CH46 1QW, UK
[2] Department of Pharmacy, University of Bath, Claverton Down, Bath, BA2 7AY
[3] Exploratory Biopharmaceutics R&D, Bristol-Myers Squibb, New Brunswick, New Jersey, USA

1 INTRODUCTION

In the pharmaceutical industry, excipients are often defined as non-functional constituents within a dosage form. Excipients are added to pharmaceutical dosage formulations for multiple reasons but most commonly they are included in order to aid the processing or to enhance stability, bioavailability, or patient acceptability of the final dosage form. Although they do not produce a medical effect, excipients have been proven to be essential in both biopharmaceutical and technical aspects. Due to their importance, interchangeability and uniformity of excipients is necessary in order to ensure consistent quality in the finished products, as well as to deal with the eventuality that a particular grade of excipient is unavailable for any reason.[1]

All raw materials, including excipients, should meet the expectations of regulatory authorities (or their delegates, for instance National Pharmacopoeias). However, pharmacopoeial testing of excipients is often based primarily on the verification of identity, purity and chemical stability with only limited testing of particle and powder physical properties. As such there are examples of excipients meeting the pharmacopoeial monograph but performing with unexpected characteristics during processing and in the final dosage form due differences in solid state characteristics.[2,3] As a result, certificate of analysis of excipients may not always provide sufficient confidence of equivalency between vendors and / or batches.

Many factors can contribute to batch-to-batch variability, from a single vendor or between multiple vendors, such as differences in raw material, manufacturing processes, storage conditions and transportation.[4] In fact, the functionality of excipients may not only depend on their intrinsic properties but also their applications and the formulation into which they are incorporated.[5] As such, it is controversial to include functionality or physical testing related to functional properties in the monograph due to the many different ways that an excipient can be used.[6] Nevertheless, a better understanding of the properties of excipients and their relationships to the functionalities can help formulators to select appropriate excipients and validate manufacturing processes accordingly, thereby improving process control and moving towards more controlled products with consistent quality in line with the FDA's Quality by Design initiative.

Microcrystalline cellulose is regarded as one of the most versatile tablet filler binders, finding wide use in both granulation and direct compression operations. Its utility is based on its low chemical reactivity, high plasticity and wide availability.[7] Microcrystalline cellulose is described as purified, partially depolymerized cellulose that occurs as a white, odorless, tasteless, crystalline powder composed of porous particles.[8] It is prepared through the hydrolysis of pulp (α-cellulose) in dilute mineral acid solutions, following which the material is filtered and then spray dried.

There has been considerable work investigating how the solid state properties of microcrystalline cellulose impact the flow and compressibility of various grades of microcrystalline cellulose[9], as well as the properties of bulk powders.[10] However, for the purposes of understanding and controlling our processes we must gain more insight into the natural variability between batches for vendors of typically used excipients.

In this paper batch-to-batch, and vendor-to-vendor variations in the solid state characteristics in multiple batches of one grade of microcrystalline cellulose from two different vendors, and the subsequent impact of these differences on processability and / or functionality is reported.

2 MATERIALS

The microcrystalline cellulose batches investigated in this study were sourced from two different vendors; Avicel PH102 (Batch No's: 70709C, 70728C, 70745C) was obtained from FMC (FMC Corp, Cork, Ireland), with two additional batches (Batch No's: P208819026, XN07818924) sourced from a second manufacturing site (FMC Corp, Newark, USA), and VivaPur 102 (Batch No's: 5610285531, 5610288045, 5610290203, 5610290506, 5610291009, 5610291512, 5610291612) obtained from JRS Pharma (JRS Pharma, Weissenborn, Germany).

3 METHODS

3.1 Specific Surface Area

Samples were analysed using a Gemini 2390A surface area analyser (Micromeritics, Norcross, USA). Samples were out-gassed for 12 hours at 100°C under nitrogen gas prior to analysis. Samples were then evacuated at a rate of 500 mmHg/min for 5 minutes and equilibrated for 5 minutes. Multipoint measurements (5 points) over the range of 0.06-0.2 p_0 were performed and the linearity within the B.E.T. range confirmed. All samples were analysed in triplicate

3.2 Inverse Gas Chromatography

The dispersive surface energy of samples was determined by inverse gas chromatography using an IGC-SMS (Surface Measurement Systems, Alperton, Middlesex, UK). Samples were packed into 0.3 m (3 mm inside diameter) silonised glass columns, plugged at either end by silonised glass wool. They were conditioned at 30°C (303 K), 0 % RH, 10 ml/min for five hours prior to analysis. The dispersive surface energy analysis was conducted by injecting a range of hydrocarbon probes; decane, nonane, octane, heptane, and hexane at 0.04 p/p_0. The polar free energy of absorption analysis was determined using a range of polar probes; acetone, acetonitrile, ethyl acetate, and ethanol at 0.04 p/p_o. The column dead time was determined using an inert probe (methane at 0.2 p/p_o). All samples were analysed in triplicate.

3.3 Particle Size Analysis

Particle size analysis was determined using a Morphologi G3 particle characterisation system (Malvern Instruments Limited, Malvern, UK). Samples were dry dispersed using the systems automated dispersion system onto a glass plate. Particle imaging was conducted using a 10x magnification lens (3.5 - 210 μm resolution range). A convexity filter was applied was applied to remove any residual aggregates or overlapping particles from the final analysis. Convexity is a measure of edge roughness and is calculated by dividing the convex hull perimeter by the measured perimeter. The convex hull perimeter can be described as the perimeter of a theoretical rubber band wrapped around the particle. Convexity values are in the range of 0 (least convex) to 1 (most convex).

3.4 True Density

True density was determined with an Accupyc 1330 helium pycnometer (Micromeritics, Norcross, USA). Samples were dried at 50°C for 12 hours prior to analysis.

3.5 Determination of Formaldehyde and Glucose

A reverse phase HPLC method with pre-column derivatization using dinitrophenylhydrazine was developed to determine trace levels of formaldehyde.[11] Kinetic studies were conducted to determine the minimum amount of dinitrophenylhydrazine and the necessary time for the derivatization reaction. Samples were prepared by suspending the excipients in 50:50 acetonitrile:water at room temperature. The suspensions were then filtered through 0.45 μm filter prior to analysis.

3.6 Angle of Repose

Samples were tested on the Geldart Mark4 Angle of Repose tester (Powder Research Ltd., UK). 100 g of each sample was poured slowly into the upper part of the chute. A motor was employed to generate a minimum degree of vibration in order to aid sample flow in the upper chute. The angle of repose (AOR) value was calculated using equation 1.

$$AOR = \tan^{-1}\left(\frac{h}{r}\right) \tag{1}$$

where h is the height of the semi-cone (mm), r average radius of the base (mm). All samples were analysed in triplicate.

3.7 Powder Compaction

Samples were compacted using a Stylcam 100R compaction simulator (MedelPharm, Bourg-en-Bresse, France). Round, flat faced compacts of 11.28 mm (equivalent to 1 cm^2 of compaction area) diameter were compressed to a target solid fraction of 0.85, which is a representative solid fraction commonly used in comparing materials. Before filling the die, the punches were manually lubricated using a 2%w/v magnesium stearate in acetone slurry to reduce tablet die ejection forces. Six compacts were produced for each sample and once ejected, the accurate weight and thickness of each tablet was measured. Yield pressure (P_y) values were determined using the Heckel analysis, using the Analis analysis package (Medelpharm, Bourg-en-Bresse, France) and the tensile strength was calculated using equation 2.

$$T = \left(\frac{2P}{\pi Dt} \right) \qquad (2)$$

where T is the tensile strength (MPa), P is the break force (N), t is the compact thickness (mm), and D is the compact diameter (mm).

3.8 Compact Hardness

Compact hardness was measured using a Schleuniger tablet tester (Dr. Schleuniger Pharmatron, Manchester, NH, USA). Six samples were tested at each condition.

3.9 Data Analysis

Data analysis was performed using Minitab 15 (Minitab, State College, PA, USA).

Table 1 *Specific surface area and dispersive surface energy results*

Vendor	Grade	Batch reference	Specific surface area [m²/g] (S.D.)	Dispersive surface energy [mJ/m²] (S.D.)
FMC Corp.	Avicel PH102	70745C	1.09 (0.02)	61.99 (0.39)
		70709C	1.16 (0.02)	66.55 (0.58)
		70728C	1.12 (0.01)	60.91 (2.10)
		XN07818924	1.05 (0.02)	58.78 (0.56)
		P208819026	1.05 (0.02)	67.83 (0.38)
JRS Pharma	Vivapur 102	5610291512	1.22 (0.00)	60.63 (0.20)
		5610291009	1.26 (0.08)	59.36 (0.31)
		5610290506	1.17 (0.02)	59.71 (0.41)
		5610290203	1.12 (0.04)	55.84 (0.25)
		5610288045	1.26 (0.05)	64.56 (0.75)
		5610291612	1.25 (0.01)	59.87 (1.00)
		5610285531	1.25 (0.01)	60.48 (0.40)

4 RESULTS AND DISCUSSION

4.1 Specific Surface Area

The specific surface area results for each of the batches of microcrystalline cellulose indicated no notable differences between the two vendors (Table 1), and intra-vendor batch-to-batch variability was also observed to be minimal. In addition, the measured surface area results were found to be equivalent to that of Avicel PH101 and PH200 grades. This is in line with previous observations, with a series of microcrystalline celluloses from Penwest Pharmaceuticals (now JRS Pharma), that surface areas were overlapping between grades, and that the interactions with a model drug (which is dependent on the surface area accessible to the compound) were equivalent between grades with nominally different particle sizes.[12] It has been shown previously that MCC particle populations consist of a mixture of 'rod like' primary particles,[13] and agglomerates, and that the proportion of these primary particles and agglomerates differs within the different grades of materials, contributing to the different bulk properties of these materials.

Therefore, as the specific surface area analysis measures the total surface area, including the surface available within the porous agglomerated particles, any relationship to the degree of agglomeration within a grade of microcrystalline cellulose is removed.

4.2 Inverse Gas Chromatography

The dispersive surface energy results for the batches of microcrystalline cellulose (Table 1) appear to be variable with notable batch-to-batch variations observed for both vendors; dispersive energy results ranged from 55.8 mJ/m^2 to 64.6 mJ/m^2 for JRS Pharma batches, and between 58.8 mJ/m^2 and 67.86 mJ/m^2 for FMC Corp. batches. These results all lie between 50 mJ/m^2 and 70 mJ/m^2, the reported values for crystalline cellulose[14] and amorphous cellulose beads,[15] respectively. The variation in the dispersive energy results for the batches of could therefore be due to differences in the availability of amorphous regions on the surface of the material, a remnant from the acid hydrolysis manufacturing process. The thermodynamically less stable amorphous materials can dominate the measured surface energy despite the fact that microcrystalline cellulose is largely crystalline[16] due to the use of a low probe concentration and the tendency of the probes to interact primarily with only the most energetic sites available.

4.3 Particle Size Analysis

Previous work by this author has proposed that the particle size distribution of this material is a mixture of both primary particles and agglomerated particles, with the proportion of the latter being numerically very low (0.5 - 2% by number) but significantly higher in volume than the primary particles.[17] Consequently, the geometric particle size measured is primarily related to the width of the particle size *distribution* rather than an increase in the mean particle size per se.

The geometric and arithmetic particle size distributions for all batches of microcrystalline cellulose investigated (Table 2) were found to be indistinguishable with the materials from each of the two vendors observed to be statistically equivalent (P=0.659). When the data is viewed in the form of an interval plot (Figure 1) it is notable that the FMC Corp. batches manufactured at the Cork plant (70709C, 70728C, and 70745C) appear to show a higher degree of batch-to-batch variability than the lots from the

Newark plant. One possible explanation for this variability could be differences in the source of the pulp used in the two countries of origin[18] or alternatively, a difference in the process control limits between the two sites; in both cases further testing of batches would be required to substantiate that the variances observed were consistent over a wider range of batches.

Table 2 *Geometric (volume based) and arithmetic (number based) particle size data*

Vendor	MCC grade	Batch reference	Geometric particle size (μm)			Arithmetic particle size (μm)		
			D_{10}	D_{50}	D_{90}	D_{10}	D_{50}	D_{90}
FMC Corp.	Avicel PH102	70745C	33.8 (4.6)	86.1 (23.5)	182.8 (24.2)	1.2 (0.1)	3.7 (1.8)	19.9 (11.3)
		70709C	43.5 (5.1)	132.1 (32.3)	255.2 (39.7)	1.2 (0.1)	3.3 (1.2)	19.1 (8.1)
		70728C	45.1 (3.3)	128.0 (30.8)	212.8 (22.6)	1.2 (0.0)	2.7 (1.0)	18.5 (15.9)
		XN07818924	39.1 (3.2)	99.1 (5.9)	209.6 (26.5)	1.3 (0.1)	4.9 (0.9)	27.5 (1.5)
		P208819026	41.7 (2.7)	100.4 (1.4)	197.8 (12.9)	1.2 (0.0)	3.1 (0.1)	20.3 (1.6)
JRS Pharma	Vivupur 102	5610291512	35.7 (1.7)	83.6 (13.3)	201.2 (18.2)	1.2 (0.0)	2.5 (0.6)	15.6 (11.5)
		5610291009	44.7 (2.9)	121.2 (7.4)	223.3 (19.8)	1.2 (0.0)	2.5 (0.3)	15.9 (6.2)
		5610290506	39.2 (1.5)	98.7 (8.4)	214.0 (17.6)	1.2 (0.0)	2.8 (0.5)	24.6 (8.5)
		5610290203	41.7 (1.1)	100.1 (6.6)	219.9 (22.2)	1.1 (0.0)	2.5 (0.2)	18.4 (7.9)
		5610288045	44.1 (2.2)	122.9 (25.2)	233.3 (20.6)	1.1 (0.0)	2.5 (0.3)	19.7 (3.7)
		5610291612	44.8 (2.8)	119.9 (9.7)	229.2 (7.7)	1.2 (0.1)	2.8 (1.2)	20.2 (18.1)
		5610285531	38.0 (3.2)	92.5 (12.4)	213.2 (22.4)	1.2 (0.1)	3.6 (2.2)	23.0 (7.2)

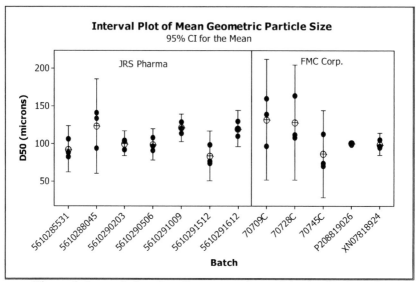

Figure 1 *Interval plot of median geometric particle size (D_{50}) results*

4.4 Angle of Repose

The angle of repose results (Table 3) indicated minimal difference between the Avicel and Vivapur batches, and no significant intra-vendor batch-to-batch variability. Despite the presence in only small fractions within the samples, the agglomerates have been shown to play a significant role in terms of flow. The results would suggest that the scale of the observed variations in the geometric particle size have little or no impact on the bulk flow properties of these samples.

Table 3 *Angle of repose, yield pressure and tensile strength results*

Vendor	Grade	Batch reference	Angle of repose (°)	Yield pressure (MPa)	Tensile strength (MPa)
FMC Corp.	Avicel PH102	70745C	33 ±2	62.1 (0.69)	8.11 (0.15)
		70709C	34 ±0	59.5 (1.15)	7.74 (0.38)
		70728C	34 ±1	59.2 (2.14)	7.71 (0.28)
		XN07818924	35 ±1	59.0 (1.08)	7.30 (0.12)
		P208819026	35 ±0	61.2 (0.61)	7.97 (0.12)
JRS Pharma	VivuPur 102	5610291512	36 ±0	64.8 (1.00)	8.27 (0.20)

5610291009	36 ±1	66.0 (1.31)	7.45 (0.13)
5610290506	37 ±1	65.7 (0.66)	7.95 (0.23)
5610290203	36 ±1	65.9 (0.42)	7.35 (0.18)
5610288045	36 ±0	66.2 (0.65)	7.95 (0.18)
5610291612	34 ±1	66.1 (0.85)	8.37 (0.36)
5610285531	37 ±1	65.8 (0.76)	8.12 (0.15)

4.5 Compressibility and Compactability

The compressibility data (Table 3) indicated a lower batch-to-batch variability for the Vivupur 102 batches, which was also found to have a slightly higher yield pressure than the Avicel PH102 batches. Microcrystalline cellulose is well-known to deform primarily by plastic deformation and its compressibility is noted to be independent of particle size.[19] One possible reason for the observed variation in compressibility is the different degree of polymerizations which can be resulted from the different wood types used for the production of microcrystalline cellulose in the two suppliers.[20]

In addition to plastic deformation, mechanical interlocking has been suggested to play a major role in the compactability of microcrystalline cellulose. It may be expected that an increase in particle size would reduce the inter-particulate bonding area and so produce weaker tablets.[21] However, no significant differences were observed between any of the samples in this study, and additional testing of samples of Avicel PH101 and PH200 also indicated no significant changes in compactability. Comparable results were also reported by Doelker et al., who observed identical compactability in unlubricated tablets produced of the three Avicel grades, particularly when taken into account of inter-batch variability.[22] However, as the number of agglomerate particles is numerically low and the primary particle distribution does not changes between grades, then the mean bonding area is unlikely to change between these grades of microcrystalline cellulose.

4.6 Determination of Formaldehyde and Glucose

The measured formaldehyde and glucose levels (Table 4) indicated a notable variance between the materials from the two vendors. The results show that the FMC Corp. batches contained significantly higher levels of glucose, and more varied and generally higher levels of formaldehyde. As a very small and reactive chemical compound, formaldehyde could cause drug product instability even at trace levels. Assuming the formulation product has a 2% drug load and a weight of 0.5 gram, 1 ppm formaldehyde corresponds to ~1% of API (molar to molar ratio). The higher levels of glucose and formaldehyde in the FMC Corp. batches could lead to an increased risk of drug degradation in formulations containing materials prone to degradation via Maillard reactions.[23,24]

Table 4 *Glucose and formaldehyde content results*

Vendor	Grade	Batch reference	Glucose (ppm)	Formaldehyde (ppm)
FMC Corp.	Avicel PH102	70709C	118.2	4.03
		70728C	201.0	0.83
		P208819026	157.7	1.88
JRS Pharma	VivuPur 102	5610285531	26.8	0.88
		5610288045	85.9	0.29
		5610291512	58.6	0.68
		5610291009	63.4	1.03
		5610290506	52.9	0.56
		5610290203	112.8	0.21
		5610291612	56.4	0.27

5 CONCLUSIONS

Although a number of variations between the two sources of material, FMC Corp. and JRS Pharma, as well as between lots from each supplier, none were found to have a profound impact of the important processability characteristics such as flowability and compactability. One notable difference between the materials obtained from the two vendors was the residual levels of glucose and formaldehyde measured; as even low levels of these species can have a significant impact on the chemical stability of materials prone to Maillard reactions.

Acknowledgements

The authors would like to thank FMC Corp. and JRS Pharma for the donation of the batches of microcrystalline cellulose used for this study. The authors would also like to thank Dr. Mridul Majumder (Pharmaterials Ltd.), Dr. Nancy Barbour, Dr. Peter Timmins, Dr. John Grosso, and Dr. Michael Leane (all Bristol-Myers Squibb) for their support during this study.

References

1 http://www.metolose.jp/e/pdf/news_20070322.pdf
2 E. Doelker, D. Mordier, H. Iten and P. Humbert-Droz, *Drug Dev. Ind. Pharm.*, 1987, **13**, 1847.
3 M. Landín, R. Martínez-Pacheco, J. Gómez-Amoza, C. Souto, A. Concheiro and R. Rowe, *Int. J. Pharm.*, 1993; **91**, 143.
4 C. Moreton, *American Pharmaceutical Review*, 2009; **12**, 28.
5 G.E. Amidon, *Physical and Mechanical Property Characterization of Powders, in: Physical Characterization of Pharmaceutical Solids*, ed. H.G. Brittain, 1995, p. 281.
6 M. Rios, *Pharm. Tech.* 2006.
7 G.K. Bolhuis and Z.T. Chowhan, *Pharmaceutical Powder Compaction Technology*, ed. G. Alderborn and C. Nyström, 1996, p. 419.

8 Handbook of Pharmaceutical Excipients, 5th Edition, Pharmaceutical Press, 2006.

9 K. Obae, H. Iijima and K. Imada, *Int. J. Pharm.,* 1999, **182**, 155.

10 T.M. Jones, *J. Pharm. Sci.*, 1968, **57**, 2015.

11 G. Huang, Y. Wu and M. Dali, 2006 HPLC Conference, poster presentation, San Francisco, CA, USA.

12 D.F. Steele, S. Edge, M.J. Tobyn, R.C. Moreton, and J.N. Staniforth, *Drug Dev. Ind. Pharm.*, 2003, **29**, 475.

13 R. Ek, G. Alderborn, C. Nyström, *Int. J. Pharm.*, 1994, **111**, 43.

14 G.M. Dorris and D.G. Gray, *J. Colloid Interf. Sci.*, 1980, **77**, 353.

15 G. Garnier and W.G. Glasser, *Polym. Eng. Sci.*, 1996, **36**, 885.

16 V. Swaminathan, J. Cobb and I. Saracovan, *Int. J. Pharm.,* 2006, **312**, 158.

17 J.F. Gamble, W-S Chiu and M. Tobyn, *Pharm. Dev. Tech.*, DOI: 10.3109/10837450.2010.495395.

18 M. Landín, R. Martínez-Pacheco, J. Gómez-Amoza, C. Souto, A. Concheiro and R. Rowe, *Int. J. Pharm.,* 1993, **91**, 123.

19 M. Celik, *Drug Dev. Ind. Pharm.,* 1992, **18**, 767.

20 G. Shlieout, K. Arnold and G. Muller, *AAPS PharmaSciTech,* 2002, **3**, article 11.

21 C. Nyström, G. Alderborn, M. Duberg and P.G. Karehill, *Drug Dev. Ind. Pharm.,* 1993, **19**, 2143.

22 E. Doelker, D. Massuelle, F. Veuillez and P. Humbert-Droz, *Drug Dev. Ind. Pharm.,* 1995, **21**, 643.

23 Y. Wu, M. Dali, A. Gupta and K. Raghavan, *Pharm. Dev. Tech.*, 2009, **14**, 556.

24 Z. Qiu, J.G. Stowell, W. Cao, K.R. Morris, S.R. Byrn and M.T. Carvajal, *J. Pharm. Sci.*, 2005, **94**, 2568.

STRUCTURAL TRANSFORMATION IN A FRICTIONAL GRANULAR SYSTEM

Q.C. Sun[1], G.H. Zhang[2], Z.W. Bi[1] and J.G. Liu[1]

[1] State Key Laboratory of Hydroscience and Engineering, Tsinghua University, China
[2] Physics Department, Beijing University of Science and Technology, China

1 INTRODUCTION

Granular materials are intrinsically athermal since their dynamics occur far from equilibrium. The majority of the theoretical considerations have been successfully applied to highly excited granular gases. For granular flows and granular solid states, they are of great engineering importance and a large number of continuum models have been presented; however, the mechanical behaviours are still rather poorly understood.[1]

Widespread interest in granular matter was aroused among physicists a decade ago. Significant progress has been made on understanding the jamming phase diagram.[2] Jamming is the physical process by which granular materials become rigid with increasing density. The phase diagram depends on temperature, load and density. A key question concerns how stability can occur when the packing fraction ϕ increases from below to above a critical value ϕ_c, for which there are just enough contacts to satisfy the conditions of mechanical stability. For frictionless soft spheres, there is a well-defined jamming transition indicated by the J point, which exhibits similarities to an (unusual) critical phase transition. In this work, the jamming/unjamming of a frictional sphere system is simulated with discrete element methods. Both mechanical and structural criteria are used to determine the J point, including the pair correlation function g(r), force-force correlations function C^n, C^t, and a position-position correlation function C^{xy}. The scaling law of the boundary pressure P with $(\phi - \phi_c)$ was examined as well.

2 METHOD AND RESULTS

Five thousand round discs in 2D are generated in a square cell of 4×4 m². To avoid crystal packings, a bidispersed distribution is used. The ratio of radii between small and large particles is 1:1.4, and the number ratio is 1:1. The constituent particle density is 2600 Kg/m³, the normal and tangential stiffness are both 1.0×10^8 N/m, and the friction coefficient is $\mu = 1 \times 10^{-4}$. The contact potential among particles is harmonic and the acceleration due to gravity is ignored. At the initial state, the radii of 5000 particle are small so that they loosely distribute in the cell. Obviously, the assembly stays in the unjammed state. The particle radius is proportionally increased to ensure the value of ϕ is enlarged at a small step of 1×10^{-4}, while keeping the radius ratio constant, until the system enters the jammed state. At each time step of small increments of ϕ, particles reach new positions after sufficient time. As shown in Figure 1, as ϕ<0.835, *P* is always close to zero; at ϕ=0.8353, *P* sharply increases to 2,017 Pa and then continuously increases. It should be noted that as ϕ increases by a step of 1×10^{-4} at 0.8356, the value of *P* suddenly drops, which indicates the phenomenon of internal stress relaxation, as shown at point A.

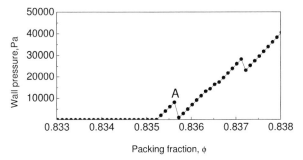

Figure 1 *Boundary pressure vs. packing fraction in the isotropically compressed system*

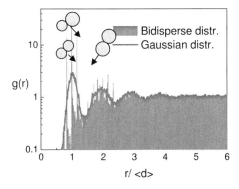

Figure 2 *Pair correlation function g(r) with the distance r. The radius r is given in the unit of the average particle diameter <d>. The height of the first peak of g(r), g1, is compared with that of the Gaussian distribution (grey curve)*

2.1 Pair-correlation function g(r)

The function $g(r)$ describes how the particle number density varies as a function of the distance r from one particular particle. For a 2D granular system, $g(r)$ is defined as:

$$g(r) = \frac{dN}{2\pi r dr} \frac{1}{\rho_0} \qquad (1)$$

where, d*N* is the particle number in the ring of r~r+dr, ρ_0 is the mean number density. For a 2D system, *N* particles distribute in the area *S*; $\rho_0 = N/S$. In this work, *dr* is 2×10^{-4}, around

1/100 of the mean diameter $<d>$. The calculated $g(r)$ distribution is shown in red in Figure 2. The grey curve is the corresponding $g(r)$ for a Gaussian size distribution of a granular assembly. It can be seen that $g(r)$ has a series of peaks for the Gaussian distribution, the first peak called g1 exists at distance $r=<d>$. However, for the bidispersed system, beside g_1 being consistent with the one obtained in the Gaussian distribution, there exist another two peaks around the first peak in the Gaussian distribution.[3, 4] This indicates two small particle contacts and two large particle contacts, respectively. It can further be seen that $g(r)$ has an oscillating shape characteristic of any disordered medium.

2.2 Force-force correlation and position-position correlation

For high fractions $\phi > \phi_c$, the decreases in $P(\phi)$ imply the transformation of one jammed state to another one. In isotropically compressed assemblies studied in this work, such drops in boundary pressure reflect the structural changes in force networks. To characterize such evolutions,[5] the force-force correlation function and position-position correlation function are introduced, and are defined as follows,

$$C^n(\phi_0,\phi) = \frac{\sum_{ij}\left|f_{ij}^n(\phi_0)\right|\left|f_{ij}^n(\phi)\right|}{\sum_{ij}\left|f_{ij}^n(\phi_0)\right|^2}, \; C^t(\phi_0,\phi) = \frac{\sum_{ij}\left|f_{ij}^t(\phi_0)\right|\left|f_{ij}^t(\phi)\right|}{\sum_{ij}\left|f_{ij}^t(\phi_0)\right|^2}, \; C^{xy}(\phi_0,\phi) = \frac{\sum_i\sqrt{x_i^2(\phi_0)+y_i^2(\phi_0)}\sqrt{x_i^2(\phi)+y_i^2(\phi)}}{\sum_i\left|x_i^2(\phi_0)+y_i^2(\phi_0)\right|} \quad (2)$$

where, taking $\phi_0 = 0.8420$ as the reference state, the sum examined over all pairs of particles (i,j) corresponds to a spatial average. f_{ij}^n and f_{ij}^t are the normal and the tangential forces; x_i and y_i are the center positions of particle i. As shown in Figure 3, after $\phi = 0.835$, the force correlation function C^n and C^t exhibit small jumps, corresponding to microslips, revealing the unusual occurrence of bursts in the reorganization of the force network. During these bursts, the energy due to the tangential interaction decreases, whereas the one due to the normal interaction increases. But, $C^{xy}(\phi_0,\phi)$ is nearly unchanged, which indicates that the geometric configuration remains unchanged. During the flat regime of $C^{xy}(\phi_0,\phi)$, $C^n(\phi_0,\phi)$ and $C^t(\phi_0,\phi)$ increase linearly with increases in the value of ϕ, which indicates that at each jammed state, the magnitudes of normal and tangential forces increase approximately linearly . At some critical value of ϕ, $P(\phi)$ rapidly increases, and $C^n(\phi_0,\phi)$, $C^t(\phi_0,\phi)$ and $C^{xy}(\phi_0,\phi)$ also change simultaneously. Although $C^n(\phi_0,\phi)$, $C^t(\phi_0,\phi)$ and $C^{xy}(\phi_0,\phi)$ change, $C^{xy}(\phi_0,\phi)$ changes with a magnitude of $\sim 5\times 10^{-5}$. This is because the particle position changes very slightly. Much larger changes are observed in $C^n(\phi_0,\phi)$ and $C^t(\phi_0,\phi)$, on the order of $\sim 10^{-1}$.

2.3 Unjamming process

The scaling of the shear modulus and bulk modulus plays a central role in connecting the disordered nature of the response to the anomalous elastic properties of systems near jamming. To understand why this disorder plays such a crucial role in the global, mechanical response of a collection of particles that act through short range interactions, consider the local motion of a packing of spherical, soft frictionless spheres under global forcing.

The unjamming process is realized by relaxing the obtained jammed states. The packing fraction becomes smaller very slowly, that is, the step of ϕ is -1×10^{-5}. It can be seen that for each jammed state, the boundary pressure P is almost linearly reduced during the unjamming process; once it has reached zero, the value of ϕ is ϕ_c. Note that for different jammed states, the ϕ_c is different, but is distributed within a narrow range from 0.83791 to 0.83820, as shown in Figure 4. It may be caused by the scale size effect, that is, when the number of particles is high, the range of variation in ϕ_c is narrow.

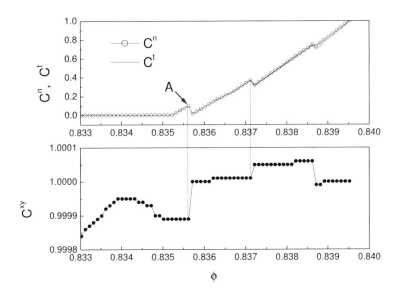

Figure 3 *Variation of C^n, C^t and C^{xy} along with increases in ϕ*

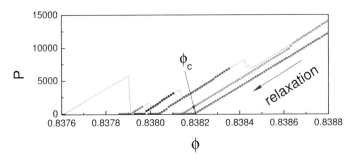

Figure 4 *Boundary pressure P with ϕ during unjamming process. For each jammed state, the value of P is almost linearly reduced during the relaxations*

Previous data has shown that bulk and shear modulii of frictionless granular systems obey scaling laws with the distance to jamming point $\Delta\phi\equiv\phi-\phi_c$, $P\sim(\phi-\phi_c)$. For frictional granular systems in this work, a similar scaling law is obtained. This is as shown in Figure 5; when $\log P \sim \log(\phi-\phi_c)$, the exponent index is 0.964. At present, there is no satisfactory physical understanding about the exponent due to the influence of friction.

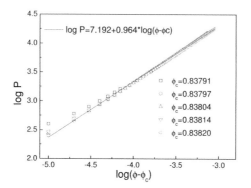

Figure 5 *Scaling laws of boundary pressure P with ($\phi-\phi_c$). The exponent index is 0.964. Symbols denote different jammed states*

3 CONCLUSIONS

The jamming/unjamming processes of the system composed of 5000 frictional disks with small friction $\mu=1\times10^{-4}$ were simulated using the discrete element method. The variation of the sidewall pressure P and the height of the first peak of the pair correlation function, g1, with packing fraction ϕ were explored. The result shows that the $P(\phi)$ curve exhibits obvious stick-slip-like behaviours, and the simultaneous adjustment of the normal force-force correlation function, tangential force-force correlation function and position-position correlation function was found during the stick-slip process. By relaxing jammed states obtained as the system underwent the compression process, we observed that the P of the different jammed states followed the same power scaling law with the distance from *J* point, $\phi-\phi_c$, $P\propto(\phi-\phi_c)^{0.964}$, although the different sidewall pressure corresponds to different value of ϕ_c.

References

1   Y.M. Jiang and M. Liu, *Granular Matter*, 2009, **11**, 139.
2   A.J. Liu and S. R. Nagel, 1998, *Nature*, **396**, 21.
3   Z. Zhang, N. Xu, D.T.N. Chen, P. Yunker, A.M. Alsayed, K.B. Aptowicz, P. Habdas, A.J. Liu, S.R. Nagel and A.G. Yodh, *Nature*, 2009, **459**, 230.
4   X. Cheng, *Phys. Rev. E*, 2010, **81**, 031301
5   M. Pica Ciamarra, E. Lippiello, C. Godano and L. de Arcangelis, *Phys. Rev. Lett.* 2010, **104**, 238001

CYCLIC LOADING CHARACTERISTICS OF A 2D GRANULAR ASSEMBLY

Z.W. Bi, Q.C. Sun, J. Fen, and J.G. Liu

State Key Laboratory of Hydroscience and Engineering, Tsinghua University, China

1 INTRODUCTION

Granular materials are collections of discrete macroscopic particles. They typically interact with each other only within short distances, mainly through collisions and contacts. The behaviour of granular materials differs from that of any other familiar forms of matter: solids, liquids, or gases. This unique property is mainly due to three factors: (i) the existence of static friction, (ii) effectively negligible temperature and, (iii) the inelastic nature of their collisions in the case of moving particles. As a consequence, granular matter shows a number of interesting features, e.g. stick-slip motion, pattern formation, or segregation.[1, 2]

A particularly intriguing phenomenon in cyclic loading systems is the so-called ratchet effect, which is introduced in order to describe the gradual accumulation of a small permanent deformation under cyclic loading. Several versions of such phenomena have been identified. The first variant occurs when the loaded sample reaches the critical state once per cycle. The mechanism is easily understood: the material would flow while it is at the critical state, giving rise to a deformation that accumulates with cycle number. However, ratcheting can also appear even when the sample never reaches the critical state. Through a series of experiments, many achievements have made and also many models have been proposed, including deformation of the limits of granular systems, the axial deformation of the three-stage evolution, loading and unloading plastic hysteresis back to evolution, fatigue damage and degradation of material properties during the evolution of micro-cracks and so on.[3,4]

However, most of previous studies on ratcheting were based on numerical simulations, but few researchers investigated the correlation between permanent displacements at the macroscopic and microscopic levels. . In this work, using our newly designed biaxial compression testing machine, we performed cyclic compression tests, from which the deformation characteristics under cyclic loading conditions were investigated, including the lateral displacement, the vertical displacement and volumetric deformation. It attempts to explore the relationship between permanent deformation mechanisms of the granular system and microscopic properties.

2 EXPERIMENTAL SET-UP AND METHODS

Figure 1 shows the diagram of the experimental set-up and some typical images. The system consists of a ternary mixture of 2500 large (diameter D = 1.0 ± 0.0025 cm), 2500 medium (0.8 ± 0.0025 cm) and 2500 small (0.6 ± 0.0025 cm) photo-elastic disks of thickness 0.3 cm that are bi-refringent under stress. The disks are placed in a Hele-Shaw cell, with dimensions L×W = 80 cm ×80 cm. The upper and left boundaries can be moved freely. The boundary can be controlled using an electro-motor under velocity control (the velocity accuracy=0.001 mm/s) and force control (the force accuracy=0.2 N). The packing fraction (φ) is defined as the ratio of the area occupied by the disks to the total cell area. The packing fraction is therefore controlled by changing the chamber area in this experiment. The global pressure is measured using a set of two sensors placed along the boundaries. Positions and displacements of individual disks as well as the stresses inferred from the photo-elastic response of the disks are recorded using a Nikon D-90 camera with 12.3 megapixel resolution. Each image captures roughly 5000 particles (roughly 50% of the total number of particles) located around the centre of the cell. The system is imaged using crossed circular polarisers. We determined the local disk contacts having a force above our experimental light intensity threshold (of about 1 N) and satisfying local force balance. The entire stress chain network satisfies physical constraints and provides a good characterization of the distribution of local forces to complement the pressure measurements.

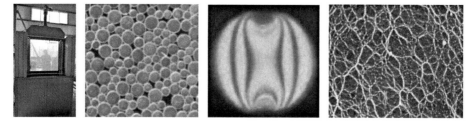

Figure 1 *Experimental set up and typical photoelastic images.*

We initially place all disks in the interrogation cell in random positions and then the sample is isotropically compressed until a force P0 =0.6 kN is reached. Then, the sample is subjected to load-unload cycles in the vertical direction as $\sigma_1 = P_0 \pm 0.6$ kN whereas $\sigma_2 = P_0$ =0.6 kN is kept constant. Loading rate remains at 0.001 kN/s.

3 RESULTS AND DISCUSSION

3.1 Evolution of axial displacement

Figure 2 shows the variation of axial force with axial displacement when the initial state is 0.6 kN and 0.8 kN, respectively. From Figure 2, it can be seen that the granular assembly experienced a significant displacement of the ratchet under the asymmetric force cycles, and ratchet displacement accumulates with the increase of loading cycles. Different from metal, the ratchet effect during the compress–tension cycle process is mainly due to apparent slip-stick. Especially in the first cycle, the force - displacement curve showed a significant fluctuation indicating slip-stick, many researchers have conducted a study of this process, but the impact of the process on the permanent deformation of granular system is rarely reported.

At the same time, it also can be seen that ratchet displacement increases with the increase of initial force. The initial force has an obvious impact on the ratchet behaviour of granular system, which need further explored. However, some literature reports that, although the initial force has a significant impact on the ratchet effect, but it has a very limited effect on the change rate of permanent displacement when the ratchet effect is accumulated to a certain extent. Due to the limited number of cycles, no further study is conducted to further explore this phenomenon.

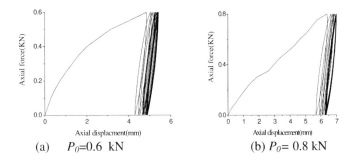

Figure 2 *Relationship between axial force and displacement*

Similar to other materials, the ratcheting displacement in the first cycle is very large, and then decreases significantly with the increase of the cycle number; after a certain cycle times, ratcheting strain rate remains essentially constant, and its value depends on the size of the initial force, i.e., the higher the initial force, the greater the permanent displacement change rate of ratcheting .

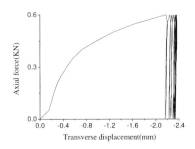

Figure 3 *The axial force as a function of the transverse displacement*

3.2 Evolution of transverse displacement

Cyclic loading under the eight different load conditions are investigated. Figure 3 shows the typical curve of axial force - lateral displacement. It can be seen that with the increase of the number of cycle, the hysteresis curve exhibit sparse - dense variation. In addition to the first cycle, the transverse permanent displacement development and recovery are very small after each loading and unloading cycle, resulting in back and forth cycle in the form of a nearly straight line for the hysteresis curve after more than 10 cycles. Meanwhile, regardless of the first hysteresis loop or the hysteresis loops afterwards, the area of hysteresis loop in transverse is smaller than that in axial. In addition, no instability is observed, which may be caused by the limited number of experiments.

3.3 Evolution of volume

As it is difficult to directly measure the volumetric change experimentally, the algebraic sum of axial displacement and transverse displacement is adopted to represent the volume change in two-dimensional simulations. Figure 4 is a typical volume change curve. It shows that the transverse displacement is between the axial displacement and volume change. It also can be seen that the average porosity decreases with the increase of the cycle number when the granular assembly is under compression. Therefore, due to the limited cycle numbers, the volume of granular system has not experienced any expansion and always under compression, and then the typical three-stage curve reported in the soil mechanics literature is not observed.

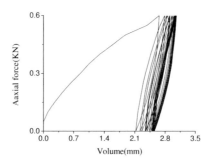

Figure 4 *A volumetric hysteresis curve*

4 OUTLOOK

Granular system in the asymmetric stress cycle will generate ratchet effect, which is obviously influenced by the initial force; the greater the initial force, the more significant the ratchet effect; but when the number of cycles reaches a certain value, the change rate of ratchet becomes independent of the initial force. The axial displacement and volume displacement are greater than the lateral displacement. In this experiment, granular system did not show significant volumetric dilatation, so there is no damage in the granular system. At the same time, regardless of experimental conditions, the displacement generated in the first cycle is the largest, which is probably caused by the apparent stick-slip process.

References

1   F. Alonso-Marroquin and H. J. Herrmann, *Phys. Rev. Lett.*, 2004, **92**, 054301.
2   S. McNamara and R.García-Rojo, *Phys. Rev. E.*, 2008, **77**, 031304.

3   F. Lekarp, U. Isacsson and A. Dawson, *J. Transportation Engineering*, 2000, **126**, 66.
4   F. Lekarp, U. Isacsson and A. Dawson, *J. Transportation Engineering*, 2000, **126**, 76.

THE IMPACT OF DRY GRANULATION ON DETERGENT POWDER PROPERTIES

A. Hart and C.-Y. Wu

School of Chemical Engineering, University of Birmingham, Birmingham, B15 2TT, UK.

1 INTRODUCTION

Detergent powders are cleaning, washing and laundering powders containing surfactants along with other ingredients, such as builders (i.e., carriers of surfactants), fillers, bleaching agents, fabric softeners, enzymes and optical brighteners.[1] Compared to liquid detergents, detergent powders have the following advantages: (i) physical and chemical stability of the ingredients, (ii) easy dispensing into the washing machines, (iii) cost effective in terms of packaging of the finished product, and (iv) cheaper for consumers. Detergent powders are generally produced through a series of unit operations, including spray drying, granulation, blending, bin filling, packing and storage.[2] Among these, granulation is a size enlargement process agglomerating fine particles together with objectives to improve the flow properties and bulk density and to reduce dusting, for which two types of granulation can be employed: wet granulation and dry granulation.

Wet granulation involves adding a liquid binder solution to powders to form granules. The liquid binder solution blends into the powder and forms bonds between the particles to lock them together when dried.[3] Using high shear wet granulation, Chateau et al.[4] examined the flowability and compressibility of detergent granules and showed that the high shear wet granulation using water as the binder improved the flowability, but reduced the compressibility of the detergent powder. This was attributed to particle agglomeration and increased density of the granules. In addition, Germana et al.[5] examined the effect of reactive binder (i.e. linear alkylbenzene sulphonic acid) and water on the wettability and the adhesive strength of the liquid bridges during detergent powder granulation using micromanipulation techniques. They found that the particle wetting is strongly dependent on the binder water content and relative humidity. In addition, a proper control in wettability during granulation process can reduce cohesion and caking in storage.

Dry granulation is an agglomeration process widely used in pharmaceutical, mineral, food and chemical industries to produce free flowing granules with improve bulk density and flowability for enhancing powder processing and handling efficiency.[6,7] This is generally achieved with roll compaction, in which the powder is compressed through two counter-rotating rollers to form compacted ribbons or flakes, which are subsequently milled to yield granules. It is expected that the produced granules have better flowing characteristics and higher bulk densities, compared to the feed powders. The impact of the operating conditions, such as roll speed, roll gap, feeding mechanism and speed, and the powder properties, such as particle size and moisture content, on the compression behaviour of powders were investigated intensively.[6,7,8,9,10] Roll compaction converts fine powders into larger granules with reduced dusting and enhanced flow properties,[11,12] but it can also lead to a decrease in mechanical properties, such as compressibility and compactibility.[13,14]

Even though roll compaction have been used widely in other industries,[6] its use in the detergent powder industry has not been reported. Therefore, it is of interest to explore the feasibility to improve flow and mechanical properties of detergent powders using roll compaction. The objective of this study is hence to examine how dry granulation (i.e., roll compaction and subsequent milling) will affect the flow and mechanical properties, such as flowability, compactibility and compressibility, of the detergent powder.

2 MATERIALS AND METHODS

A Professional Ariel® regular powder (Procter & Gamble, UK) was used as the model material (feed powder). The true densities of the feed powder and produced granules were measured using a helium gas pycnometer (AccuPyc™ II 1340, Micrometritics®, Germany) and are 2.0147 ± 0.01 g/cm^3 and 1.9678 ± 0.02 g/cm^3, respectively.

The flow properties were measured using a Schulze ring shear cell tester (RST-XS, Dietmar Schulze, Wolfenbuttel, Germany). The powder was first pre-sheared under a consolidation stress of 10 kPa until a steady consolidated state is reached. The samples were then sheared under four normal stresses: 8.0, 6.0, 4.0, and 2.0 kPa to construct a yield locus. The ring shear cell tester was also used to measure the wall friction angle φ_w and adhesion between the sample and a smooth surface.

Size distributions of the feed powder and granules were determined using the dry powder laser diffraction particle size analyser (Sympatec, Germany). Sampling was performed with the aid of a spinning riffler (Retsch PT1000, UK). The powders were dispersed with a pressure of 5 bars to break up agglomerates and to prevent attrition of primary particles so that the particle size distribution can be determined accurately. Each test was run in triplicate.

Roll compaction experiments were performed using an instrumented lab-scale roll compactor developed at the University of Birmingham (see Wu *et al.*[6] for details). The detergent powder was compacted through roll gaps of 1.0 and 1.2 mm with roll speeds ranging from 0.5 to 5.0 rpm. The compacted ribbons were then milled using a cutting mill (Retsch SM100, Germany) with a screen opening size of 10 mm.

Uniaxial compression tests were performed with maximum compression forces ranging from 1 to 3 kN at a constant compression rate of 60 mm/min using a universal mechanical testing machine (Zwick-Roell Z030, Germany). The feed powder and granules were compressed in a die of 13 mm in diameter with flat-faced punches. The dimensions of the compacted detergent tablets were measured with a digital calliper. The tablets were then diametrically compressed at a compression speed of 0.5 mm/min in order to determine the tensile strength. The maximum force required to break the tablet was recorded and tensile strength, σ_t, was determined as follows[15]

$$\sigma_t = \frac{2F}{\pi dt}$$
(1)

where F is the maximum force during diametrical compression, d is the diameter of the detergent tablet and t is the tablet thickness.

3 RESULTS AND DISCUSSION

3.1 Flowability

Flowability of a powder can be characterised using the flow function, ff_c, defined as the ratio of the consolidation stress to unconfined yield strength. According to Jenike,[16] the larger the value of flow function, the better the powder will flow. The variations of the flow function with the normal stress for the feed powder and produced granules are shown in Figure 1. It is clear that the flow function generally decreases as the applied normal stress increases. This indicates that flowability reduces when a powder is subjected to a higher compression stress. In actual powder handling processes, the stresses acting on the powder could be much higher than the stresses employed in the shear cell test. This implies that processing problems related to poor powder flowability, such as caking and arching, are more likely to occur when the power is subjected to enduring high consolidation stresses. In addition, the feed powder has a slightly higher value of flow function, compared to the produced granules, i.e., the feed powder has a slightly better flowability than the granules. This indicates that dry granulation does not improve the flowability for this particular detergent formulation.

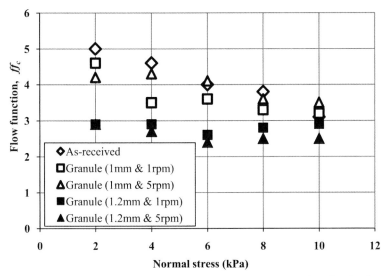

Figure 1 *Variation of flow function with normal stress for the feed powder and produced granules.*

The feed powder and produced granules were also sieved into different size cut, and the flow functions for the samples in various size ranges were measured using the ring shear cell tester and presented in Figure 2. It can be seen that the produced detergent granules generally have poorer flowability than the feed powder even in the same size range. It is believed that this is due to the change in the shapes of particles and granules, i.e. from round (as-received) to angular (granules), resulting in the increase in internal friction angle (Figure 3) and decrease in flowability.

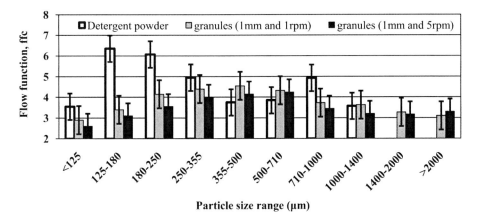

Figure 2 *Variation of flow functions with particle size for the feed powder and produced granules in various size ranges.*

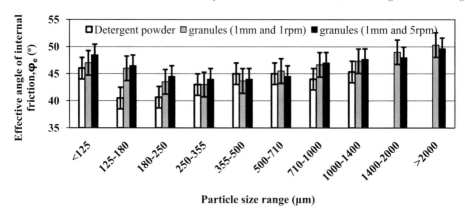

Figure 3   *Variation of effective internal friction angle with particle size for the feed powder and produced granules.*

3.2 Compressibility

Compressibility is referred to as the densification (*i.e.* reduction in volume) of a powder bed in response to an applied pressure, which can be analysed using Heckel equation:

$$ln\left(\frac{1}{1-D}\right) = kP + A \tag{2}$$

where D is the relative density, P is the applied stress, k and A are Heckel constants. The value of k indicates the compressibility (*i.e.* the change in powder volume at a given compression pressure) and a large value of k corresponds to a better compressibility. A is related to particle rearrangement during compression.

The compressibility profiles for the feed powder and produced granules are shown in Figure 4, in which all samples were compressed to a maximum compression force of 3 kN. The corresponding k values are presented in Figure 5. It is clear from Figs 4 & 5 that, at the same compression pressure, the feed powder experiences larger volume reduction than the granules, indicating that there is a reduction in compressibility once the powder is roll-compacted. Furthermore, the compressibility of the granules reduces as the roll speed increases and as the roll gap decreases. This is consistent with the results for microcrystalline cellulose[13] and pseudopolymorphic forms of theophlylline.[17]

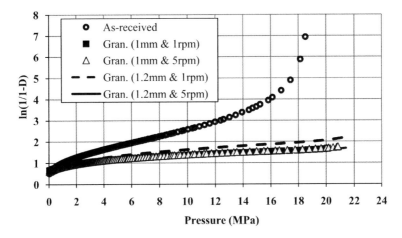

Figure 4 *Heckel plots for the feed powder and produced granules (Maximum compression force: 3kN).*

3.3 Compactibility

Compactibility is the ability of a powder to cohere into a compact, which is generally characterised using the tensile strength of tablets produced at a given maximum compression pressure (*i.e.* compaction pressure). The compactibility of the feed powder and produced granules are presented in Figure 6, in which the relationships between tensile strength and compaction pressure for various samples considered are shown. It is clear that the higher the compaction pressure, the higher the tensile strength of the compacts, which is in broad agreement with the data for various powders reported in the literature.[13, 18-21]

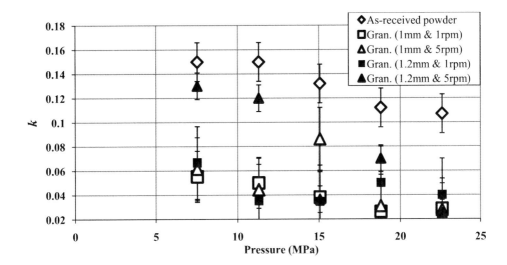

Figure 5 *Heckel constant k as a function of compression pressure.*

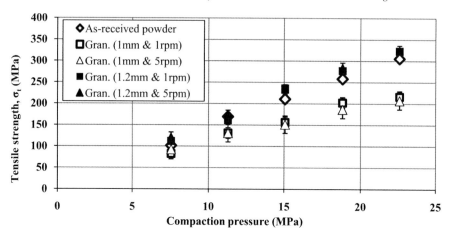

Figure 6 *Tensile strength as a function of the compaction pressure for the feed powder and produced granules.*

In addition, the tablets made from granules produced at a smaller roll gap (1.0 mm) have lower tensile strengths, compared to the feed powder and granules produced at a larger roll gap (1.2 mm). This is attributed to the reduced compressibility (Figs 4 & 5), in other words, as the compressibility is reduced, tablets with lower solid fractions were produced at the same compaction pressure, which leads to the reduction in tensile strength.

The data presented in Figure 6 were re-plotted against the porosity of the tablets for the feed powder and produced granules in Figure 7. It is interesting to find that, for all tablets produced, the variation of tensile strength with porosity coalescence into a single master curve that can be approximated using the Ryshkewitch-Duckworth equation[18]

$$ln\left(\frac{\sigma_t}{\sigma_o}\right) = -k_m\varepsilon \qquad (3)$$

where ε is the porosity, σ_t is the tensile strength and σ_o *is* the tensile strength at $\varepsilon = 0$.

This clearly demonstrates that the tablet porosity is the dominant factors for tensile strength. The difference in tablet tensile strength for various tablets produced with the same compaction pressure (Figure 6) is primarily due to the variation in compressibility.

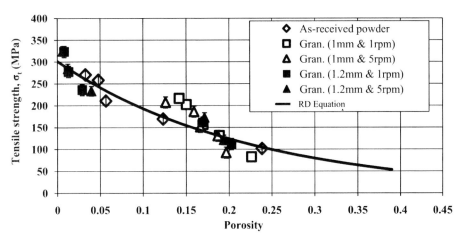

Figure 7 *Variation of tensile strength with the tablet porosity.*

4 CONCLUSIONS

The effect of dry granulation on flowability, compressibility and compactibility of a detergent powder was investigated. The powder was roll compacted to ribbons that were subsequently milled into granules. The flowability, compressibility and compactibility of the feed powder and granules produced under various roll compaction conditions were determined. It has also been found that roll compaction does not improve the flow properties for this particular powder. It has been also shown that the granules produced through dry granulation can lose their compressibility and compactibility, especially when they are produced through roll compaction at a narrow roll gap (hence high compaction pressure). Nevertheless, for all samples considered, the tablet tensile strength is a unique function of tablet porosity, which can be approximated using the Ryshkewitch-Duckworth equation, indicating that porosity is the dominant controlling factor in determining the tensile strength of powder compacts.

Although dry granulation has been used for size-enlargement in powder handling industries and it has been shown that it can improve the flowability for some powders, for more complicated formulations, like the finished detergent powder considered in this study, it may have adverse effects on flowability and compressibility. It is recognised that dry granulation can reduce the powder compactibility, which is further demonstrated in this study with a commercial detergent powder. Nonetheless, it has been shown that the porosity is a critical factor determining the tensile strength of powder compacts.

References

1 Y.,Yangxin, J. Zhao and E. B. Andrew, *Chinese Journal of Chemical Engineering*, 2008, **16** (4), 517-527
2 U. Zoller and S. Paul, *Handbook of Detergent, Part F: Production*, CRC Press, 2009, 323-355

3 D. M. Tousey, *Pharmaceutical Technology Tableting & Granulation*, 2002, www.dipharma.com/The_Granulation_Process_101.pdf, accessed 28/08/2011.

4 M. Chateau, G. Laurence, S. Yannick and F. Jacques, *Powder Technology*, 2005, **157**, 191-198.

5 S. Germana, S. Stefaan, B. Judith and C. Brendan, *Powder Technology*, 2009, **189**, 385-393.

6 C.-Y. Wu, W.-L. Hung, A.M. Miguelez-Moran, B. Gururajan and J.P.K. Seville. *Int. J. Pharm.*, 2010, **391**, 90-97

7 A.M. Miguelez-Moran, C.-Y. Wu and J.P.K. Seville. *Int. J. Pharm.*, 2008, **362**, 52-59

8 A. Y. Yusof, S. C. Andrew and B. J. Brain, *Chem. Eng. Sci.*, 2005, **60**, 3919-3931.

9 G. Bindhumadhavan, J.P.K. Seville, M.J. Adams, R.W. Greenwood and S. Fitzpatrick, *Chem. Eng. Sci.*, 2005, **60**, 3891-3897

10 R.F. Mansa, R.H. Bridson, R.W. Greenwood, H. Barker and J.P.K. Seville, *Powder Technology*, 2008, **181**, 217-225.

11 P.C. Knight, *Powder technology*, 2001, **119**, 14-25

12 K.J. Prescott and A.R. Barnum, *Pharmaceutical Technology*, 2000, http://www.jenike.com/Techpapers/on-powder-flowability.pdf. Accessed 19 June, 2010. pp. 60-84

13 M.G. Herting and P. Kleinebudde, *Int. J. Pharm.*, 2007, **338**, 110-118.

14 C.C. Sun and M.W. Himmelspach, *J. Pharm. Sci.*, 2006, **95**, 200-206.

15 J.T. Fell and J.M. Newton, *J. Pharm. Sci.*, 1970, **59**, 688-691.

16 A.W. Jenike, *Bulletin of the University of Utah, 1970, No. 123*, University of Utah, Salt Lake City.

17 E. Hadzovic, B. Gabriele, H. Seherzada, E. K. Silvia and L. Hans, *Int. J. Pharm.*, 2010, **396**, 53-62.

18 C.-Y. Wu, S. Best, C. Bentham, B. Hancock and W. Bonfield, *Euro. J. Pharm. Sci.*, 2005, **25**, 331-336.

19 K. C. Tye, C. Sun and A. E. Gregory, *J. Pharm. Sci.*, 2005, **94** (3), 465-472.

20 Y. Zhang, Y. Law and S. Chakrabarti, *AAPS Pharm. Sci. Tech.*, 2003, **4** (4), Article 62.

21 M.J. Sonnergaard, *Euro. J. Pharm. and Biopharm.*, 2006, **63**, 270-277.

Processing

THE EFFECT OF CARRIER PARTICLE SIZE ON ADHESION, CONTENT UNIFORMITY AND INHALATION PERFORMANCE OF BUDESONIDE USING DRY POWDER INHALERS

W. Kaialy,[1, 2] H. Larhrib[1] and A. Nokhodchi[1]

[1]Chemistry and Drug Delivery Group, Medway School of Pharmacy, University of Kent, Kent ME4 4TB, UK
[2]Pharmaceutics and Pharmaceutical Technology Department, University of Damascus, Damascus 30621, Syria

1 INTRODUCTION

Dry Powder Inhalers (DPIs) are the result of the development of two technologies: powder technology and device technology. Particle deposition in the respiratory tract is affected by many aerosol particle properties such as particle size, shape, density, charge, and hygroscopicity.[1] In particular, particle size is of great importance as it is known that particle-particle interactions within DPI formulations are related to van der Waals forces. Therefore, particle size is the most important physical property and design variable of a DPI formulation. Several studies were reported on the effect of drug particle size on DPI performance, showing that the preferred drug particle size is between 1-5 μm.[2] However, in literature, the effect of carrier particle size distribution (PSD) on drug aerosolisation efficiency has received less attention and reported in dissimilar manner.[3] Nevertheless, it should be noted that there is rare studies aimed to show the effect of carrier particle size as a single variable factor on DPI performance. In this study, the effect of lactose particle size distribution on budesonide adhesion, content uniformity and *in vitro* aerosolisation performance was investigated.

2 METHODS AND RESULTS

2.1 Micromeretic, solid state, and flow properties

Commercial α-lactose monohydrate powder was sieved to obtain different lactose samples with different size fractions as follows: A (90-125 μm), B (63-90 μm), C (63-45 μm), D (20-45 μm) and E (< 20 μm). Laser diffraction and scanning electron microscope observations showed that different lactose powders have considerably different size distributions and different surface topographies (Figure 1). Span values indicated narrower size distributions for lactose powders with smaller volume mean diameter (VMD) (r^2=0.9792, Span= 0.417 ln (VMD) +2.6386). Fine particle lactose ($FPL_{<10μm}$), which was reported to have dominating effect on DPI performance,[4] was absent in sample A, B, and C whereas sample D and E contain 5.2±0.2% and 36.8±2.4% of $FPL_{<10\mu m}$, respectively. Higher specific surface area (SSA_v) was obtained for lactose samples with smaller volume mean diameter, higher span (linear, r^2=0.9723), and higher fine particle lactose (linear, r^2=0.9736).

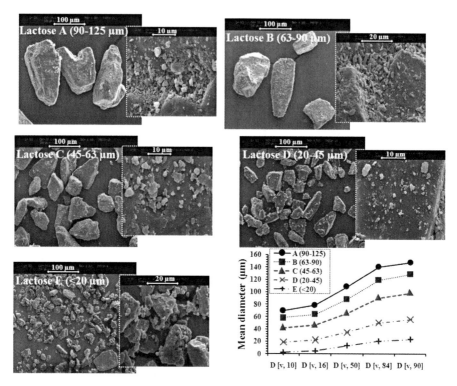

Figure 1   *Scanning electron micrographs (SEM) and PSD of different lactose samples.*

Particle shape image analysis showed that lactose powders with smaller size have higher shape factor, higher surface factor, smaller roughness (Figure 2a), higher roundness, and higher angularity (Figure 2b). This indicates smoother surface and higher degree of shape regularity for lactose particles with smaller size.

Differential scanning calorimeter was employed to characterise the crystalline nature of all lactose samples (Figure 2). All lactose samples showed the typical thermal curve of α-lactose monohydrate consisting of two distinctive endothermic peaks at about 148° C and 219°C and one smaller exothermic peak at about 175 °C corresponding to dehydration of crystalline hydrate water, melting of α-lactose, and crystallisation of amorphous lactose.[5]

Lactose samples with smaller size showed larger exothermic peak at about 175 °C, which is indicative of higher amorphous content. By calculating % amorphous content of different lactose samples,[6] linear relationship was established showing that the lactose samples with higher volume mean diameter have smaller amorphous content (Figure 3).

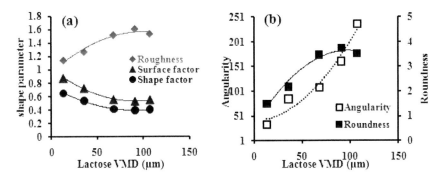

Figure 2 *Roughness, surface factor, and shape factor (a); angularity and roundness (b) for different lactose samples.*

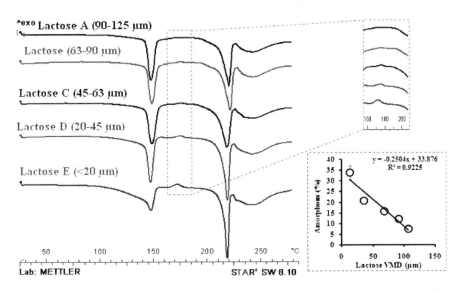

Figure 3 *Differential scanning calorimetry thermograms of different lactose samples.*

True density (D_{true}) measurements provided by helium pycnometery showed that lactose powders with higher volume mean diameter have smaller true density (linear, r^2=0.9932). On the other hand, lactose powders with higher volume mean diameter showed higher bulk density (D_b) (linear, $r^2=0.8943$) and higher tap density (D_t) (linear, $r^2=0.8244$). Lactose powders with smaller VMD showed higher Carr's index values (linear, $r^2=0.9177$) (indicating poorer flow properties) and higher porosity (linear, $r^2=0.914$).

2.2 Uniformity, adhesion, and *in vitro* aerosolisation performance assessments

Five different formulations were prepared by blending micronized budesonide (median diameter=3.2±0.2 μm) with different lactose samples (A, B, C, D, and E) in a ratio of 1:67.5 w/w in Turbula[TM] mixer for 30 min. From each blend, at least seven samples were collected randomly for quantification of budesonide content using High Performance Liquid Chromatography. All blends showed similar drug content potency ($p<0.05$); however lactose particles with smaller VMD produced higher coefficient of variation (CV%) of budesonide indicating reduced drug content homogeneity (Figure 4a). This can be attributed to poorer flowability and wider size distribution for lactose powders with smaller particle size as shown previously. In fact, direct linear relationships were obtained when plotting coefficient of variation of budesonide against lactose Carr's index ($r^2=0.9275$) or span ($r^2=0.8776$).

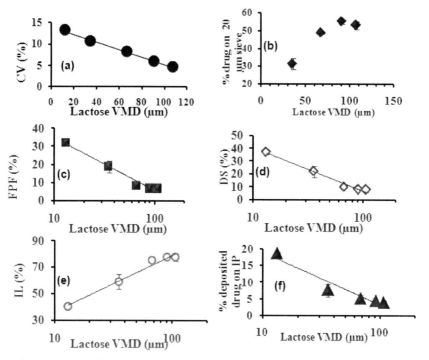

Figure 4 *Relationship between lactose VMD and budesonide CV (a), amount remained on top of 20 μm sieve (b), FPF (c), DS (d), IL (e), and amount deposited on IP (f).*

Air jet sieving was employed to evaluate drug-carrier adhesion properties for all formulations. Less amounts of drug remained on top of the 20 μm sieve was obtained for the lactose particles with smaller VMD (Figure 4b) indicating weaker drug-carrier adhesion. This could be, in part, attributed to higher collision and friction forces during mixing process for lactose powders with higher VMD, which act as adhesive forces.

In vitro aerosolisation performance of different formulations was analysed using Multi Stage Liquid Impinger (MSLI) attached to Aerolizer® inhaler device. The results showed that despite using the same batch of budesonide in all formulations; budesonide aerodynamic particle size was dependent on lactose particle size. Lactose particles with smaller size produced budesonide particles with smaller aerodynamic size upon inhalation. Higher fine particle fraction (FPF) (Figure 4c), higher dispersibility (Figure 4d), and smaller impaction loss (IL) (Figure 5e) of budesonide were obtained for lactose particles with smaller volume mean diameter indicating improved aerosolisation performance. This could be attributed to smaller drug-carrier adhesion for lactose powders with smaller volume mean diameter (Figure 4b) and consequently improved drug-carrier detachment efficiency upon inhalation. However, it was noticed that the smaller the lactose volume mean diameter, the higher the amounts of budesonide deposited on throat (IP) (Figure 4f), which is disadvantageous in terms of increased potential local and/or systemic side effects of budesonide.

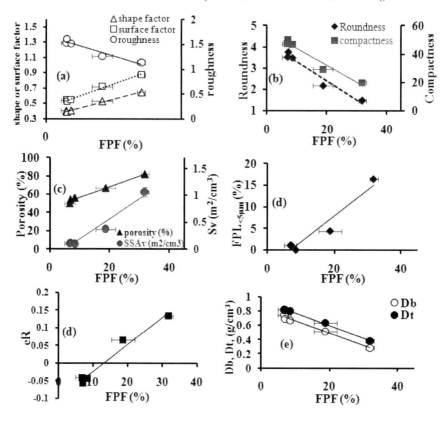

Figure 5 *FPF of budesonide in relation to lactose shape factor, surface factor, roughness, roundness, compactness, porosity, SSAv, FPL$_{<5\mu m}$, e$_R$, D$_b$, and D$_t$.*

Direct linear relationship (r^2=0.9822) was established between amounts of drug deposited on throat and fine particle lactose (FPL$_{<10\mu m}$) (figure not shown). Lactose powders with poorer flowability produced higher amounts of budesonide remained in inhaler device and deposited on throat. It was assumed that high powder cohesiveness for lactose D and E (as indicated by higher Carr's index values) could not be totally overcome during inhalation process, leading to the formation of aggregates remained in the inhaler device and/or deposited on throat. The smaller the volume mean diameter of lactose, the smaller the amounts of budesonide deposited on MSLI stage 1 and the higher the amounts of drug deposited on MSLI stages 2, 3, 4, and filter. Figure 5 shows that higher fine particle fraction of budesonide was obtained when lactose powders with smaller roughness, higher shape factor, higher surface factor, higher roundness, higher compactness, higher porosity, higher specific surface area, higher fine particle lactose, higher simplified shape factor, smaller bulk density, and smaller tap density were used. This indicates that the aerosolisation performance is better when carrier particles with smoother surface, regular shape, higher surface area, higher content of fines and smaller bulk and tap density are used.

3 CONCLUSIONS

This study showed that the smaller the carrier size the better the drug aerosolisation efficiency. However, the use of carrier powders with smaller particle size is disadvantageous in terms of reduced dose homogeneity, higher potential of side effects possibility, and reduced formulation stability and flowability.

Abbreviations

CI:	Carr's index;	CV:	coefficient of variation;
D_b:	bulk density;	DPI:	dry powder inhaler;
DS:	dispersibility;	D_t:	tap density;
D_{true}:	true density;	e_R:	simplified shape factor;
FPF:	fine particle fraction;	FPL:	fine particle lactose;
IL:	impaction loss;	IP:	induction port;
MMAD:	mass median aerodynamic diameter;	MSLI:	Multi Stage Liquid Impinger;
PSD:	particle size distribution;	SEM:	scanning electron microscope;
SSAv:	specific surface area;	VMD:	volume mean diameter.

References

1   A.J. Hickey and T.B. Martonen, *Pharm. Res.*, 1996, **10**, 1-7.
2   P. Zanen, L.T. Go and J.W.J. Lammers, *Int. J. Pharm.*, 1994, **107**, 211-217.
3   H. Weyhers, W. Mehnert and R.H. Müller, *Pharmazeutische Industrie*, 1996, **58**, 354-357.
4   X.M. Zeng, G.P. Martin, C. Marriott and J. Pritchard, *J. Pharm. Sci.*, 2001, **90**, 1424-1434.
5   A. Gombas, P. Szabo-Revesz, M. Kata, G. Regdon and I. Erös. *J. Therm. Anal. Calorim.*, 2002, **68**, 503–10.
6   A. Saleki-Gerhardt, C. Ahlneck and G. Zografi, *Int. J. Pharm.*, 1994, **101**, 237–47.

THE CONTROL OF MICRODISPENSED MEAN DOSE OF INHALAC®70 BY VARYING THE TIME OF VIBRATION IN AN ACOUSTIC MICRO-DOSING SYSTEM

Z. Li, L. Pan, P.N. Balani and S. Yang

School of Engineering Sciences, University of Southampton, SO17 1BJ, UK
Email: s.yang@soton.ac.uk

1 INTRODUCTION

In a pharmaceutical research and development set up, only few grams of new chemical entities as synthesized powders are available. Micro to nanograms of these chemical entities undergo high throughput screening for further identification as potential drug candidates. The screening time of these candidates governs the development time of drug products.[1] Currently, dispensing is carried out using either vacuum aspiration or feeding using screw/ augur or vibration into the vial. However, drawbacks such as time, high capital cost, constant re-checking by weighing balances and lower fill yield due to fine powders loss at the filter are exposed especially during dispensing of cohesive and adhesive micron-sized inhalation powders.[2] The dose content uniformity, surface morphology of the drug and carrier, drug-carrier ratios are important aspects in dry powder inhaler (DPI) formulations as it affects the efficacy of the inhaled drug.[3,4]

Commonly employed pneumatic aspiration-ejection or volumetric powder dispensing methods do provide a solution with dispensing of small dose masses at higher accuracy and higher manufacturing speeds. However, dependence on powder properties such as packing density with low reproducibility in dispensing of cohesive pharmaceutical powders such as starch limits their usage. Vibratory devices utilizing vibration energy could overcome these limitations, wherein a piezoelectric layer on surface of a hopper controls the flow of the material as the powder is dispensed using a mechanical valve. However, mechanical valve blockage due to the constant movement is the major limitation of these devices.[5]

Ultrasonic vibration controlled micro-dosing has evolved as a feasible alternative in accurate dispensing of fine powders. The use of acoustic vibrations from a piezoelectric disc aids in breaking the agglomerated powder clots by applying a distributed and continuous force on the powders. As a result, these vibrations when switched on, can initiate the flow of powder from a fine glass nozzle without using a mechanical stopper. On switching off the vibrations, particle-particle and particle-wall frictions lead to formation of domes causing powder flow arrest in the nozzle.[6,7] The efficient functioning of the device is dependent on inter-play between different process variables such as nozzle diameter, water depth in the tank, voltage amplitude, voltage frequency and time of vibrations.[5] Time of vibration has been found to be a critical process variable controlling the dispensing of mean dose mass of fine materials such as tungsten carbide, steel tool and glass beads.[8] The prolonged time of vibration has been found to increase the mean dose mass, which has been linear in case of some powders such as tungsten carbide.

The present study looks into the effect of time of vibration of these devices in controlling the micro-dispensed mean dose masses of low dose, micron-sized inhalation grade lactose, InhaLac®70. InhaLac®70 is sieved crystalline form of α-lactose monohydrate and is widely used as a carrier in DPIs.[9]

2 EXPERIMENTAL METHOD

The experimental facility as reported earlier[5] consists of a computer, a D/A card (NI 6733, National Instruments Corporation Ltd. Berkshire, UK), a power amplifier (50w, Sonic Systems Ltd, Somerset, UK), a glass nozzle placed in a water tank with a piezoelectric ceramic disc (SPZT-4 A3544C-W, MPI Co., Switzerland) attached at the bottom. The piezoelectric disc excited by the high frequency signal (40 kHz) transmits the vibration to the nozzle through water. The upper section of the water tank functions as a hopper for the powder sample. The experiments were conducted by varying vibration time from 0.1 to 1 s at fixed signal amplitude of 5V and nozzle size of 0.8 mm. Vibrations are provided in the form of a pulse with a time interval of 6 s. The time intervals are included to allow stabilization of the microbalance. The total time was defined as the circle time as shown in Figure 1 with the next pulse induced on completion of one circle time. The dispensed dose mass was recorded and verified by the microbalance. The staircase curve dispensing pattern could be uniform if the curve is forwarded leading to a constant gradient of the curve. The experiments were carried out in a closed chamber dried using silica gel a desiccators at room temperature. As morphology and size distribution govern the behaviour of particles in the nozzle, morphological analysis of InhaLac®70 (Meggle GmbH Wasserburg, Germany) was performed using scanning electron microscopy (SEM). The particle size distribution was determine using Motic Images plus v.2.0 software.[10] A high speed camera (FASTCAM SA5, Photron USA, Inc) operating at 6000 fps was used to capture images during dispensing of InhaLac®70 particles from the nozzle at different time of vibration.

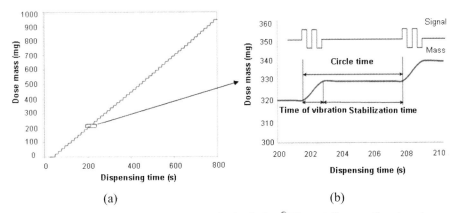

(a)   (b)

Figure 1 (a) *Accumulated dose dispensed of InhaLac®70 at different vibration time and (b) the schematics of total time involved in dispensing of a single dose*

3 RESULTS

Figure 2 shows the mean dose masses of InhaLac®70 dispensed at different time of vibration, fixed signal amplitude of 5 V and nozzle size of 0.8 mm. On initiation of ultrasonic vibrations in the range of 0.1 – 1 s, a linear increase in dispensed mean dose mass of InhaLac®70 was noted. The particle flow ceased as the vibration was stopped. A variable dosage in the range of 1.68 mg to 15.12 mg could be achieved. In addition, dispensing speeds or flow rates calculated by dividing mean dose by respective pulse time of vibration was plotted in Figure 1. A stable dispensing speed of 15.75 ± 0.75 mg/s with 4.8% RSD was achieved by varying time of vibration. Matsusaka and co-workers[11] have reported similar flow behaviour on micro-feeding of fly-ash spherical fine particles with particle size of 15 μm from a vibrating capillary tube. The adhesive fine particles adhering to the inner wall of the capillary tube act as a micro-vibrating layer of particles which avoid contact of the larger or small agglomerates with the wall of the capillary tube. This micro-vibration behaviour lowers the frictional stress between the inner powder and the wall, causing the agglomerates to fall under gravitational force. On increasing the frequency of vibrations, these micro-vibrating particles undergo increase activity as observed in the video microscopic images collected during micro-feeding. This further reduces the frictional stress between the particles and the wall surface leading to an increase in flow rate. The flow rate however saturates beyond a certain upper limit of frequency, as the activity of micro-vibration of particles no longer increases at higher frequencies. The similar team of researchers[12] earlier studied the micro-feeding of fine particles under the influence of ultrasonic vibrations and found that changes in powder flow rate was directly proportional to particle properties such as packing fraction, velocity, particle density and diameter of the capillary tube. As increases in time of vibration affect the velocity of moving particles in the nozzle, it can have a similar influence on powder flow rate until a certain point as observed in our study.

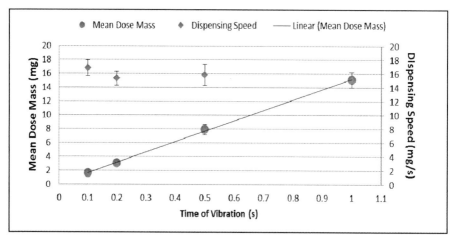

Figure 2   *Mean dose mass and dispensing speed of InhaLac®70 at different time of vibration.*

Powders with large particle size, high density and good flow are dispensed uniformly using acoustic micro-dosing systems.[5] These properties along with ultrasonic vibrations aid in achieving a balance between compaction (due to particle-particle interactions) and dilation (due to ultrasonic vibration) enabling uniform dosing from the nozzle. On initiation of ultrasonic vibrations, these powders extrudes as rods or cluster or discrete particles at the tip and break off into uniform doses beyond the orifice. The arching behaviour preferred for stable microfeeding in these systems has been reported earlier.[5] Powders such as tungsten carbide with these characteristics show increase in dispensed mean dose masses at prolonged time of vibration.[5,8] The relationship is almost linear above 10 ms of vibration with formation of integral doses producing weight changes at equal steps. This allows stable micro-dosing.[5] Further, variations in mean dose mass distribution decreases at higher time of vibration.[8]

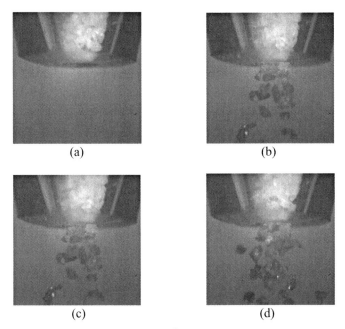

(a) (b) (c) (d)

Figure 3 *High speed images of InhaLac®70 from (a) static nozzle tip at 0 s and dispensing as discrete particles at time of vibration of (b) 0.1 s, (c) 0.5 s, and (d) 1 s*

In the present study, less cohesive InhaLac®70 with an agglomerated large-size broadly distributed particles (150 – 200 μm) and low fine content fell as discrete particles as visualized in the photographic images captured at different time of vibration using high speed camera (Figure 3). Particle size range was in accordance to the earlier report with no change in morphology.[9] The flow halted on stopping the vibrations indicating that powders did form arches in the nozzle. The arches formed are constantly disturbed by vibrations enabling flow and recreated by jamming. Janda and team in their extensive study on understanding the jamming behaviour of powder particles in a vibrated hopper found that an increase in vibrations could lead to a gradual decrease in the hopper orifice diameter and in the absence of vibrations the flow could be continuous for a certain period followed by a complete blockage.[13] It was also reported that flow rate increases at higher vibrational amplitudes until a certain time of vibration is reached where the mean flow rate is similar. A regular breakage of arches and jamming with a well defined time spent in jammed state lead to a linear rise in dispensed mean dose mass of InhaLac®70.

The role of vibration could further be explained by a circular arc model derived by Matchett using vibration induced flow of a contained material in a conical hopper.[14] Under the influence of vibrations, two limiting conditions were considered, the push mode where vibrations caused the material to move towards the cone axis leading to plastic compression of the materials due to passive stress, and the pull mode in which the vibrations causes the wall of the hopper to release stress over the contained material and lead to a switch in stress state from passive to active which causes the material to flow. Since the InhaLac®70 particles are contained in the nozzle and could be termed as "contained material" suggests the applicability of this model to explain the flow behaviour in the present study. The stable dispensing speed suggests that switch from passive to active state occurred at time of vibration of 0.1 s and remained constant at ≥0.2s.

4. CONCLUSIONS

The present study shows that varying the time of vibration of acoustic micro-dosing system at a fixed voltage of 5 V in a 0.8 mm nozzle size leads to dispensing of mean dose mass of 1.68 – 15.12 mg of InhaLac®70. The dispensed mass fell in the form of discrete particles with a linear correlation to time vibration suggesting that an accurate and stable dosing process of InhaLac®70 could be modulated depending on different time of vibration. The findings suggest that time of vibration is an important process parameter in acoustic micro-dosing system which enables controlling dosage of inhalable micron-sized lactose.

References

1   Y. Jiang, S. Matsusaka, H. Masuda and Y. Qian, *Powder Technol.,* 2009, **188,** 242.
2   J.H. Bell, P.S. Hartley and J.S.G. Cox, *J. Pharm. Sci.,* 1971, **60,** 1559.
3   E.M. Littringer, A. Mescher, H. Schröttner, P. Walzel and N. A. Urbanetz, *Sci Pharm.,* 2010, **78,** 672.
4   X.M. Zeng, G. P. Martin, C. Marriott and J. Pritchard, *J. Pharm. Pharmacol.,* 2000, **52,** 1211.
5   X. Lu, S.F. Yang and J. R.G. Evans, *J. Phys. D: Appl. Phys.,* 2006, **39,** 2444.
6   S.F. Yang and J.R.G. Evans, *Powder Technol.,* 2004, **142,** 219.
7   S.F. Yang and J.R.G. Evans, *Phil. Mag.,* 2005, **85,** 1089.
8   X. Lu, S.F. Yang and J.R.G. Evans, *Powder Technol.,* 2007, **175,** 63.
9   G. Pilcer and K. Amighi, *Int. J.Pharm.,* 2010, **392,** 1.
10   N.D. Cekić, S.D. Savić, J. Milić, M.M. Savić, Ž. Jović and M. Maleševic, *Drug Del.,* 2007, **14,** 483.
11   S. Matsusaka, K. Yamamoto and H. Masuda, *Adv. Powder Technol.* 1996, **7,** 141.
12   S. Matsusaka, M. Urakawa and H. Masuda, *Adv. Powder Technol.* 1995, **6,** 283.
13   A. Janda, D. Maza, A. Garcimartín, E. Kolb, J. Lanuza and E. Clément, *EPL.,* 2009, **87,** 24002.
14   A.J. Matchett, *Chem. Engg. Res. Des.,* 2004, **82,** 85.

THE EFFECT OF SIGNAL AMPLITUDE OF AN ULTRASONIC VIBRATION CONTROLLED METERING DEVICE ON MICRO-DISPENSING OF INHALATION GRADE LACTOSE

Z. Li, L. Pan, P.N. Balani and S. Yang

School of Engineering Sciences, University of Southampton, SO17 1BJ, UK
Email: s.yang@soton.ac.uk

1 INTRODUCTION

Dry powder micro-dispensing can be used in many fields, such as solid freeform fabrication, drug delivery and pharmaceutical screening.[1-4] Micro-dispensing of powders also provide a large landscape in terms of formulation designs for high throughput experimentation in the pharmaceutical industry. Micro to milligrams of powders are metered and dispensed for application in combinatorial chemistry of drug development.[5] Conventional dispensing methods have limitations such as high capital cost, time consumption, operational complexity, constant use of weighing balances and lower fill yield due to fine powders loss at the filter. The process is more challenging when dispensing cohesive and adhesive micron-sized inhalation powders. The mass content uniformity of each dose in such formulations is especially crucial as it assures consistent therapeutic benefits in patient.[6]

The volumetric dosing of an active pharmaceutical ingredient (API) in a dry powder inhaler (DPI) requires mixing with coarser carrier particles such as lactose to provide sufficient flowability. There is a strong dependence on the inter-particulate forces between API and carrier to ensure good powder uniformity and DPI efficacy. Other factors such as surface morphology of the carrier also govern efficient delivery of the API.[7] A higher carrier ratio warrants an accurate metering and dispensing in order to ensure dose uniformity.[8]

With a need to develop high speed and accurate dispensing machines for manufacturing of pharmaceutical products, several methods of powder metering and dispensing have been explored. The most common of these are pneumatic and volumetric dispensing methods which are simpler techniques and can operate at higher manufacturing speeds with low dose variation. In pneumatic conveying the powder in the groove of a rotating disc is pneumatically conveyed by venturi to cyclone for mixing and deposition. The motor speed controls the powder flow rate. However, fine powders are drawn away by cyclone.[9] Volumetric powder dispensing devices can usually dispense powder with higher accuracy (less than 1% variation in dose mass deviation) and relatively high speed (3 to 11g/s) but are very sensitive to any change of packing density. However, these methods are inaccurate in dispensing cohesive and "sticky" pharmaceutical powders such as starch.[10] As suitable alternative, vibratory devices have been explored; this provided advantages such as improved flow and could dispense dry powders through a mechanical valve from a conical hopper. Vibrations were generated using a flexing piezoelectric layer on hopper surface which transmits a vibratory signal to the dry powder facilitating fluidic flow.[11] The device needs a mechanical valve to close and open the hopper outlet. The need of a valve serves as a major limitation as it gets jammed due to constant movement during dispensing.[10]

Yang and co-workers have developed a dispensing device to eliminate the need of mechanical stopper. The device can "print" dry powders analogue to a desktop drop-on-demand inkjet printer. The powder is dispensed (without a mechanical stopper) by application of an acoustic energy driven system, with glass capillary as funnel and delicate computer control system. The vibrations from a piezoelectric disc can precisely initiate and halt the flow of powder from a fine nozzle with no mechanical stopper. Once the ultrasonic vibration is switched off, the powder flow arrest is brought about by the formation of domes in the capillary due to wall-particle and particle-particle frictions. The dosage can be controlled by parameters of the system including diameter of the nozzle and the intensity of the ultrasonic vibration, as well as powder properties and environment.[12,13] Yang and team have been able to accurately dispense fine materials such as tungsten carbide, steel tool and glass beads. Additionally, the uniformity of the mean dose mass dispensed was affected by signal amplitude of the vibration pulse and increasing the signal voltage amplitude was found to widen the dose mass distribution.[14]

With preliminary studies indicating viability of this method in accurately dispensing dry powders, the application of this method is further explored in dispensing of micron-sized lactose powder widely used in DPIs. Thus, the purpose of this work is to study the effect of vibration signal amplitude as one of the process variables on the dispensing of mean dose mass of micron-sized inhalation grade lactose, InhaLac®70. InhaLac®70 is sieved crystalline lactose and is used as a carrier in DPIs.[15]

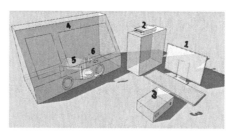

1. Computer control platform; 2. D/A card; 3. Power amplifier; 4.Glove box; 5. 7 place microbalance; 6. Dispensing facility

Figure 1 *Experimental arrangement of the ultrasonic vibration controlled micro-dispensing system*

2 EXPERIMENTAL METHOD

The experimental facility, shown in Figure 1, comprises a computer, D/A Card (National Instruments Corporation Ltd. Berkshire, UK), a power amplifier (50w, Sonic Systems Ltd, Somerset, UK), a glove box operating at room temperature containing a microbalance (2100 mg ± 0.1 µg, Sartorius AG, Germany) and a dispensing facility comprising of a glass nozzle (made from a capillary tube) placed in a purpose-built water tank with a piezoelectric ceramic ring (SPZT-4 A3544C-W, MPI Co., Switzerland), attached to the bottom of the tank with an adhesive commonly used in ultrasonic cleaning tank construction (9340 GRAY Hysol Epoxi-Patch Structural Adhesive, DEXTER Co., Seabrook, USA). The piezoelectric transducer excited by the high frequency signal (40 kHz) transmits the vibration to the capillary through water. The upper section of the water tank functions as a hopper and stored the feed powder. The experiments were conducted at different signal amplitude voltages (2.5 -5 V) using a 0.8 mm glass nozzle. Silica gel beads were placed in the glove box to provide dry atmospheric conditions. The microbalance verified and recorded the dose mass and the data collected is transmitted to the computer via RS232 serial-port. Morphological analysis of InhaLac®70 was performed using scanning electron microscopy (SEM). The particle size distribution was determined by analyzing collected SEM images using Motic Images plus v.2.0 software.[16] A high speed camera (FASTCAM SA5, Photron USA, Inc) operating at 6000 fps was used to capture image during dispensing from the nozzle.

3 RESULTS

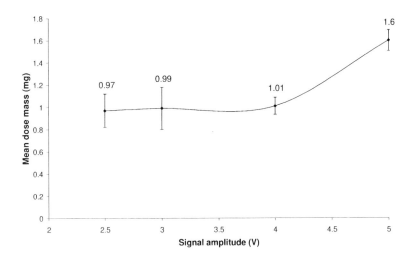

Figure 2 *Mean dose mass of InhaLac®70 dispensed at different signal amplitudes*

Figure 2 shows the effect of increases in signal amplitude on the dispensed mean dose mass of InhaLac®70. At a pulse vibration of 0.1 s and signal amplitude of 2.5 V, an initial mean dose mass of 0.97 could be dispensed. No change in dispensed mean dose mass was evident on increasing the signal amplitude to 4V. However, a significant increase in mean dose mass could be seen on vibrating at amplitude of 5V. The discrete doses were uniformly dispensed at each amplitude as indicated by the run charts in Figure 3a-d where mass of 10 individual doses are recorded. The powder flow rates were calculated at each amplitude by dividing mean dose by pulse vibration of 0.1s. Mean flow rates of 9.7, 9.9, 10.1 and 16.1 mg/s at vibrational amplitudes of 2.5, 3, 4 and 5 V were obtained. The changes in powder flow rate could be correlated to similar changes in the dispensed mean dose mass.

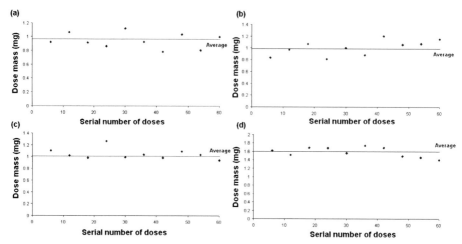

Figure 3   *Dose mass distribution of InhaLac®70 dispensed at (a) 2.5 V (b) 3V (c) 4V (d) 5 V*

The maximum force (F_b) produced by a piezoelectric block in a vibrated system is influenced by the signal amplitude as explained by Loverich.[17] This was exemplified by Lu and co-workers while studying the effect of different parameters on the ultrasonic microfeeding behaviour of fine powders.[18] The authors found that H13 tool steel fine powder showed linear increases in mean dose mass dispensed, while titanium dioxide showed a similar change in mean dose mass as InhaLac®70 on increasing the signal amplitude from 3-6 V under square wave form actuation. In addition, an upper limit of voltage amplitude was defined beyond which an irregular dose mass was dispensed indicated from the width of the error bars. The irregular dosing pattern is also evident in dispensing of InhaLac®70 from the wide error bars seen in Figure 2. However, the wide error bars are found when the amplitude is lower (≤ 3 V), while a more regular dosing behaviour with lesser deviation is noted beyond 3 V. One explanation may be that lower vibration amplitudes reduces the occurring of those events which bring the system into blockage as explained by Janda and team.[19] This meant that the flow rate was higher and regular at 5 V as seen at in Figure 3d.

The success of a microfeeding system has been found to be dependent on the behaviour of the powder in the capillary tube. The arching behaviour is the most preferred one as it allows controlled dispensing of the powder in the form of either as columnar rods, discrete or cluster of particles.[18] InhaLac®70 as shown in Figure 4a appears as large agglomerated crystals with a smooth surface having low fine content. The particle size distribution was in the range of $100 - 200$ µm which tallied with the earlier report.[15]

These powder properties along with an angle of repose of $31°$ suggest that InhaLac®70 is free flowing. The flow as discrete particles as seen in high speed camera image in Figure 4b confirms the arching behaviour of InhaLac®70 particles while being dispensed. The stability of the arch in vibrated conical hopper systems has been explained using a model derived by Matchett.[20] Under the influence of vibrations, two limiting conditions i.e. push mode and pull mode has been considered which governs the flow behaviour of contained materials through small outlet diameters. The push mode leads to compression of the material by vibrated wall section, while pull mode releases the stress upon the contained material and causes a switch in stress orientation from passive to active stress state. The active stress state causes the arch to collapse and material to flow. The switch to active stress should be quick and the flow should be established before it reverts back to the push mode. The achievement of active stress state also requires the vibrations to cover large portion of the hopper and should last for larger part of the vibration cycle for the flow to be reliable. In order to achieve this, larger amplitude or duration is needed. Subsequently, the collapse of the arch generates failure in all the sections of the hopper affecting the flow rate of the contained powder, leading to a possible flow even through small outlet diameters.

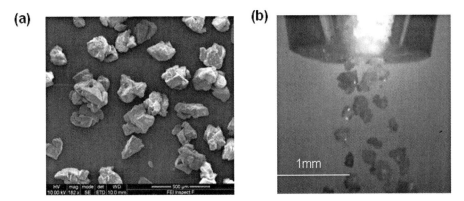

Figure 4   (a) *Scanning electron photomicrograph of InhaLac$^{®}$70 (b) High speed imageof InhaLac$^{®}$70 falling as dispersed discrete particles from the capillary tube*

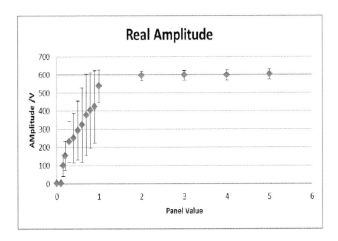

Figure 5 *A plot of real values of amplitude measured using an oscilloscope and compared with signal amplitude displayed on panel of ultrasonic amplifier*

In the present study, the significant increase in flow rate and mean dose mass at 5 V could be due to the switch to an active stress state. However, the switch could not be correlated to changes in amplitude as the real amplitude remained constant above 2 V (Figure 5) suggesting saturation of the output from the ultrasonic power amplifier.

4. CONCLUSIONS

The present work shows that flow rate and dosing of InhaLac®70 through a 0.8 mm nozzle at 0.1 s pulse of vibration could be controlled by modulating the signal voltage of an ultrasonic vibration controlled microdispensing system. The discrete flow of particles suggested that the external applied voltage contributed to forming of stable arches with irregular flow at ≤ 3 V, while the arch collapse leading to a uniform flow and increase in mean dose mass of InhaLac®70 signal amplitudes of > 4 V. These results provide a strong basis for exploring the application of an ultrasonic controlled microdispensing system as a feasible method in dispensing of inhalation grade micron-sized pharmaceutical materials.

References

1   K. Seppälä, J. Heinämäki, J. Hatara, L. Seppälä and J. Yliruusi, *AAPS Pharmscitech*, 2010, **11**, 402.
2   N.R. Kane, B. Broce, J. Gonzalez-Zugasti, W.P. Lewis, M. Lequesne and A.V. Lemmo, *J. Assoc. Lab. Autom.*, 2004, **9**, 219.
3   Y. Yamada, T. Akita, A. Ueda, H. Shioyama and T. Kobayashi, *Rev. Sci. Instrum.*, 2005, **76**, 062226-1.
4   S.F. Yang and J.R.G. Evans, *Mater. Sci. Eng. A.*, 2004, **379**, 359.
5   S.L. Morissette, Ö. Almarsson, M.L. Peterson, J.F. Remenar, M.J. Read, A.V. Lemmo,
    S. Ellis, M.J. Cima and C.R. Gardner, *Adv. Drug Del. Rev.*, 2004, **56**, 275.
6   J.H. Bell, P.S. Hartley and J.S.G. Cox, *J. Pharm. Sci.*, 1971, **60**, 1559.
7   E.M. Littringer, A. Mescher, H. Schröttner, P. Walzel and N. A. Urbanetz, *Sci Pharm.*, 2010, **78**, 672.
8   X.M. Zeng, G. P. Martin, C. Marriott and J. Pritchard, *J. Pharm. Pharmacol.*, 2000, **52**, 1211.
9   S.F. Yang and J.R.G. Evans, *Mater Sci Eng .*, 2004, **A379**, 351.
10 S.F.Yang, *WO/2011/020862.*, 2011.
11 T.M.Crowder, A.J.Hickey, V. Boekestein, *US-A-2004/0050860.*, 2004.
12 S.F. Yang and J.R.G. Evans, *Powder Technol.*, 2004, **142**, 219.
13 S.F. Yang and J.R.G. Evans, *Phil. Mag.*, 2005, **85**, 1089.
14 X. Lu, S.F. Yang and J.R.G. Evans, *Powder Technol.*, 2007, **175**, 63.
15 G. Pilcer and K. Amighi, *Int. J.Pharm.*, 2010, **392**, 1.
16 N.D. Cekić, S.D. Savić, J. Milić, M.M. Savić, Ž. Jović and M. Maleševic, *Drug Del.*, 2007, **14**, 483.
17 J.J. Loverich, *PhD Dissertation*, 2004, Pennsylvania State University.
18 X. Lu, S.F. Yang and J. R.G. Evans, *J. Phys. D: Appl. Phys.*, 2006, **39**, 2444.
19 A. Janda, D. Maza, A. Garcimartín, E. Kolb, J. Lanuza and E. Clément, *EPL.*, 2009, **87**, 24002.
20 A.J. Matchett, *Chem. Engg. Res. Des.*, 2004, **82**, 85.

CAPSULE FILLING PERFORMANCE OF POWDERED FORMULATIONS IN RELATION TO FLOW CHARACTERISTICS

T. Freeman, [1] V. Moolchandani, [2] S.W. Hoag [2] and X. Fu [1]

[1] Freeman Technology Ltd., Malvern, WR13 6LE, UK
   E-mail:xiaowei.fu@freemantech.co.uk
[2] School of Pharmacy, University of Maryland, Baltimore, MD21201, USA

1 INTRODUCTION

Among pharmaceutical solid dosage forms, the capsule is second only to the compressed tablet in terms of popularity because it is easy to swallow, does not require a coating step and can, uniquely, be subjected to multiple filling operations for different materials into the same capsule.

The introduction of powders into capsules can be divided into three sub processes:
1. Die filling – powders enter into dosing disc cavities;
2. Compaction – powders are consolidated to form slugs;
3. Ejection – slugs are ejected into capsule body.

Thus the capsule filling operation requires the powder to be sufficiently free flowing to fill the dosing cavity but at the same time, the powder needs to have sufficient strength to form a robust slug which can then be efficiently ejected intact into the capsule shell.

There are some studies in the literature which investigate the relationship between the formulations, operation variables and filling efficiency.[1-7] However, these studies rely on single parameters such as angle of repose or Carr's Index to describe the powder flow properties. This approach is neither systematic nor representative of a real processing environment. Correlations identified from one formulation are often not valid for other formulations.

A powder's flowability is complex and its effect on capsule filling is difficult to predict, but there are dynamic, bulk and shear properties that in combination can provide increased understanding of how powders perform in such processes. This study examines the ejection force and the dosing weight variation during a capsule filling process for twelve different formulations and shows how the results correlate with the powder characteristics obtained using a Powder Rheometer.

2 MATERIALS AND METHODS

Twelve formulations based on four grades of lactose (Pharmatose 110M, AL, DCL11 and DCL15) were selected for this study. Each lactose was lubricated (0.5wt.% MgSt), and the lubricated mixture was also combined with 10wt.% and 40wt.% Acetaminophen (APAP) to generate 3 formulations based on each grade of lactose.

The particle size distributions of lactose grades were analysed using a Mastersizer (Malvern Instruments Ltd., UK). The particle density was measured using an AccuPyc 1330 Pycnometer (Micromeritics Instrument Corporation, USA). The results are tabulated in Table 1. The powder flow characteristics were measured using an FT4 Powder Rheometer (Freeman Technology Ltd., UK).

Table 1 *Physical properties of different grades of lactose*

Lactose Type	Materials Description	Mean Particle Size VMD, D(4,3), um	Bulk Density g/ml	Particle Density, g/ml
110M	α - Lactose monohydrate	129	0.75	1.54
AL	Anhydrous lactose	152	0.66	1.58
DCL11	Spray dried α - Lactose monohydrate	105	0.68	1.54
DCL15	Granulated α - Lactose monohydrate	100	0.52	1.55

A Harro Höfliger-KFM/3 (Harro Höfliger GmbH, Germany) dosing disc machine was employed in this study and a schematic of the filling and tamping processes is shown in Figure 1. At station 1, powders enter the cavity under both gravity and force as the pins push through the powder bed above and a slug is formed. When the tamping process is complete, the pins withdraw, the dosing plate rotates, and the slugs are positioned in the next tamping station. The tamping process is repeated until the cavities are completely filled at station 5. Finally, the slugs are positioned over empty capsule bodies at station 6, where they are ejected by transfer pistons.

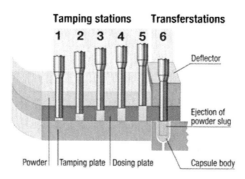

Figure 1 *Schematic drawing of the dosing disc machine (Harro Höfliger-KFM/3)*

The formulations listed in Table 2 were encapsulated using a Size 1 piston under 150N, at 35rpm rotation speed and with a powder bed height of 40mm. In this study, three tamping stations were employed and two tamping strokes were given at each station. The slug weight and slug ejection force were monitored during capsule filling. The ejection force, fill weight and coefficient of variation (%CV) of fill weight for the twelve formulations are shown in Table 2.

Table 2 *Capsule filling performance of 12 formulations (n=10)*

	APAP content (%)	Plug Wt (mg)	% CV (Plug wt)	Ejection Force (N)
	0	452	1.58	25
110M	10	418	4.50	30
	40	424	4.30	38
	0	444	0.41	30
AL	10	448	3.00	35
	40	392	2.80	44
	0	366	0.52	25
DCL11	10	328	6.30	35
	40	375	5.10	35
	0	371	0.48	25
DCL15	10	355	3.60	37
	40	351	2.80	43

3 RESULTS AND DISCUSSIONS

3.1 Flow characteristics vs. APAP content

The powder flow characteristics measured by the FT4 are shown in Figure 2. The results clearly show that those formulations containing a higher percentage of APAP consistently generate higher values for Specific Energy, Compressibility and Cohesion, and lower values for Aeration Ratio – all suggesting higher cohesivity under different environmental stresses [2]. The Permeability of the formulations decreases with increasing APAP percentage.

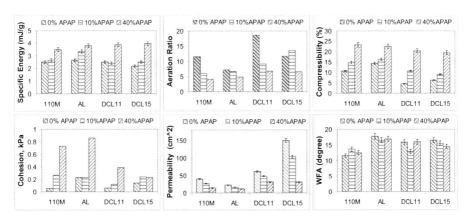

Figure 2 *The representative powder flow characteristics of the twelve formulations*

Wall Friction Angles (WFA, 1.2μm 316L stainless steel Wall Friction Disc) of all formulations were determined. No clear trends were observed between the APAP content and WFA values. For the same grade of lactose, the WFA values for the different formulations are similar with variation of less than 1 degree (7%).

3.2 Ejection force vs. flow characteristics

Empirical approaches are usually preferred in real applications, in which a correlation of flow characteristics to the operational parameters is desired.

The relationships between each formulation's APAP percentage, powder flow characteristics and ejection force are established by correlation and shown in Figure 3. The results clearly reveal that those formulations containing higher percentage of APAP are consistently associated with a higher ejection force.

The ejection force is dependent on radial force produced by decompression/relaxation of the slug, and coefficient of friction between the slug and wall. As the Wall Friction Angles for the formulations with different APAP content are very similar, this suggests that coefficient of wall friction is not the influential variable for these formulations. The ejection force instead correlates with powder cohesivity, indicating that this is the dominant factor which influences the radial force, and therefore the ejection force.

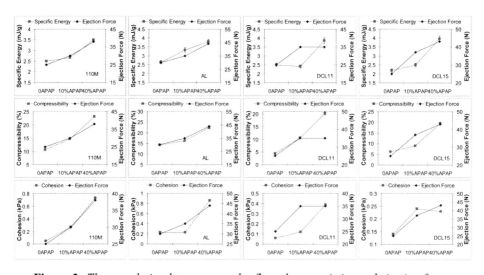

Figure 3 *The correlation between powder flow characteristics and ejection force: from top to bottom – Specific Energy, Compressibility and Cohesion*

3.3 Weight variability vs. flow characteristics

The formulations containing APAP show an increased variation in capsule weight. Furthermore, the formulations containing 10wt.% APAP, which show moderate cohesivity, exhibit consistently higher weight variation than formulations containing 40wt.% APAP (Figure 4).

Slug segments are actually a composite of powder that enters the dosing disc cavities during disc rotation and additional powder that enters the cavities during tamping as the pins push through the powder bed over the dosing disc (Figure 5). Therefore the overall weight variability of the capsule is a balance between the variability in the two different filling modes, gravity filling and forced filling.

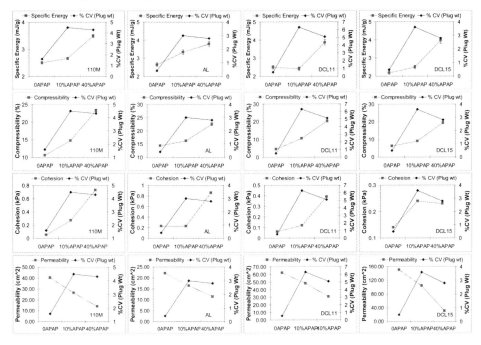

Figure 4 *The correlation between powder flow characteristics and capsule weight variability: from top to bottom – Specific Energy, Compressibility, Cohesion and Permeability*

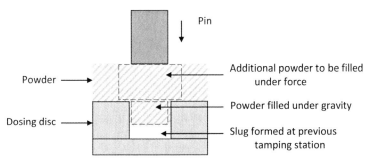

Figure 5 *Schematic drawing of the two filling stages during the capsule filling using the dosing disc machine*

As the dosing disc rotates to each tamping station, the powder enters the dosing disc cavity under centrifugal force and gravity. During this gravity-filling process, both the ability of the powder to flow into the die and the interaction between the powders and entrapped air are dominant factors [3, 4]. Previous research has shown that efficient gravity filling requires low cohesivity and high Permeability of the powders [5, 6], which suggests that the formulations without APAP will tend to have the best filling efficiency during this process.

This is not the case during the forced filling, where efficient flow is favoured by powders with lower Permeability and higher cohesivity [7]. When the pins push through the powder bed, those powders with higher cohesivity and lower Permeability – containing more entrained air, have reduced bed stiffness and allow improved flow under force. This would suggest that the formulations with 40% APAP have the best filling efficiency during this forced filling process.

The variation in performance of the same formulation during the two different sub-processes may explain why formulations with moderate cohesivity and permeability have the highest weight variability in this capsule filling process as their flow properties are not conducive with either of the sub-processes. Furthermore, this may also explain why correlations identified from formulations based on one excipient are often not valid for formulations based on a different excipient [1], as the weight variability in each individual sub process varies with the formulation's flow characteristics.

4 CONCLUSIONS

The results clearly show that those formulations containing higher percentage of APAP consistently have higher cohesivity and lower Permeability, and are associated with higher ejection force. This strong correlation indicates that the powder cohesivity is a dominate factor in relation to the ejection force.

The formulations containing APAP show an increased variation in capsule weight. Furthermore, the formulations containing 10wt.% APAP, which show moderate cohesivity, exhibit consistently higher weight variation than formulations containing 40wt.% APAP (Figure 4). This is likely to be due to the variation in filling performance of the same formulations during the two different filling sub-processes, gravity filling and forced filling. Thus the importance of understanding the precise relationship between powder characteristics and sub-process is vital.

The results shown in this study suggest that the inclination to create the moderately cohesive formulations must be scrutinised in the context of actual capsule filling performance. An approach based on multiple characteristics is more reliable for us to understanding and predict the system behaviour.

References

1   P.K. Heda, K. Muteba and L.L. Augsburger, *AAPS PharmSci*, 2002, **4(3)** E17.
2   R. Freeman, *Powder Technology*, 2007, **174**, 25.
3   C.Y. Wu and A. C. F. Cocks, *Powder Metallurgy*, 2004, **47 (2)**, 127.
4   C.Y. Wu, L. Dihoru and A. C. F. Cocks, *Powder Technology*, 2003, **134**, 24.
5   R. Freeman and X. Fu, *Powder Metallurgy*, 2008, **51 (3)**, 196
6   R. Freeman, J. Cooke and L.C.R. Schneider, 2007, *PARTEC*, Nuremburg, Germany
7   R. Freeman and X. Fu, 2008, *Particulate Systems Analysis*, Stratford-upon-Avon, UK

THE ROLE OF NOZZLE SIZE OF AN ACOUSTIC VIBRATION CONTROLLED MICRO-DOSING SYSTEM ON MICRO-DISPENSING OF INHALAC®70

Z. Li, L. Pan, P. N. Balani and S. Yang

School of Engineering Sciences, University of Southampton, SO17 1BJ, UK
Email: s.yang@soton.ac.uk

1 INTRODUCTION

The process of synthesizing new chemical entities and identifying them as potential drug candidates leads to a limited availability of few grams of synthesized powders.[1] This warrants the need to develop techniques to dispense precisely microgram or even nanogram quantities of these powders for its application in high-throughput screening enabling faster drug development. Current dispensing technologies are tedious with limitations such as high capital cost, operational complexity, constant use of weighing balances and lower fill yield due to fine powders loss at the filter. The challenge is more acute when dispensing cohesive and adhesive micron-sized inhalation powders.[2] The volumetric dosing of an active pharmaceutical ingredient (API) in a dry powder inhaler (DPI) requires uniform mixing with coarser carrier particles such as lactose to provide sufficient flowability. Additionally, inter-particulate forces between API and carrier affects powder uniformity and eventually DPI efficacy.[3] The dose content uniformity in DPIs is especially crucial as it assures consistent therapeutic benefits in patient. Surface morphology and higher ratio of the carrier also affect efficient delivery of the API.[3,4]

Pneumatic and volumetric dispensing are the most common and simpler techniques commonly employed for dispensing as they provide suitable operational feasibility at higher manufacturing speeds with low dose variation. However, these methods depend on powder packing density and are not suitable for dispensing of cohesive pharmaceutical powders such as starch.[5] Although, use of vibratory devices provide advantages such as improved flow with the use of a mechanical valve from a conical hopper, the valve could get jammed during the dispensing process.[6]

Microfeeding by a computer-controlled micro-dosing system has emerged as a promising and feasible alternative, where in the acoustic vibrations from a piezoelectric disc can initiate and halt the flow of powder from a fine conical glass nozzle with no mechanical stopper. Flow of the powder is ceased as vibration is switched off. The powder flow arrest is brought about by the formation of domes in the nozzle due to wall-particle and particle-particle frictions.[7,8] Previously, Deming and Mehring examined the flow of granular solids from conical bins and derived an equation for the flow rate, which was restricted by the conical angle.[9] Similarly, other researchers such as Newton et al[10], Brown and Hawksley[11], derived flow rate equations for solids from circular orifices, however they were only limited to certain orifice sizes and particle sizes much larger than pharmaceutical materials. The micro-dosing device developed by Yang and co-workers has been able to accurately dispense precise dose masses of fine materials such as tungsten carbide, steel tool and glass beads.[12] The nozzle size has been found to be one of the important process parameters affecting the dispensing of mean dose mass. The selection of the nozzle size is crucial as it is dependent on the powder properties which are essential in controlling dispensed dose mass in order to avoid blockage (if too narrow) and free flowing (if too large). The mean dose mass range could be altered using a range of nozzle diameters.[6]

The purpose of present research work is to explore the role of nozzle diameter of an acoustic vibration controlled micro-dosing device on micro-dispensing of low dose, micron-sized inhalation grade lactose, InhaLac®70. InhaLac®70 is sieved crystalline lactose and is widely used as a carrier in DPIs.[13]

2 EXPERIMENTAL METHOD

The experimental facility, shown in Figure 1, comprises a computer, D/A Card (National Instruments Corporation Ltd. Berkshire, UK), a power amplifier (50w, Sonic Systems Ltd, Somerset, UK), a dry glove box (dried using silica gel as desiccants) operating at room temperature; containing a microbalance (2100 mg ± 0.1 µg, Sartorius AG, Germany), a glass nozzle placed in a purpose-built water tank with a piezoelectric ceramic disc (SPZT-4 A3544C-W, MPI Co., Switzerland) which is attached to the bottom of the tank with an adhesive commonly used in ultrasonic cleaning tank construction (9340 GRAY Hysol Epoxi-Patch Structural Adhesive, DEXTER Co., Seabrook, USA). The piezoelectric transducer excited by the high frequency signal (40 kHz) transmits the vibration to the nozzle through water. The upper section of the water tank functions as a hopper for the powder sample. The experiments were conducted using different nozzle sizes (0.6 – 0.9 mm) at fixed signal amplitude of 5V and pulse vibrations of 0.1 s. The microbalance verified and recorded the dose mass. Morphological analysis of InhaLac®70 was performed using scanning electron microscopy (SEM). The particle size distribution was determined by analysing collected SEM images using Motic Images plus v.2.0 software.[14] A high speed camera (FASTCAM SA5, Photron USA, Inc) operating at 6000 fps was used to capture image during dispensing from the nozzle.

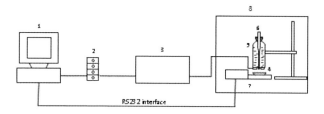

1. Computer control platform; 2. D/A card; 3. Power amplifier; 4.Piezoelectric disc; 5. Water tank; 6. Nozzle; 7. Microbalance , 8. Glove box.

Figure 1 *Experimental arrangement of the acoustic vibration controlled micro-dosing system.*

3 RESULTS

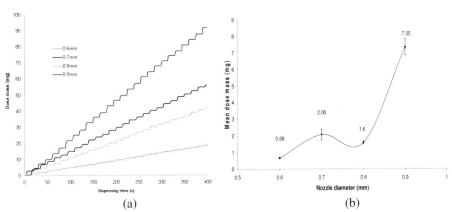

(a)   (b)

Figure 2 *(a) Accumulated doses dispensed as a function of time elapsed and (b) mean dose mass of InhaLac®70 using different nozzle sizes*

Figures 2a and 2b shows the accumulated doses and mean dose mass of InhaLac®70 dispensed using different nozzle sizes under ultrasonic pulse vibration of 0.1 s and signal amplitude of 5 V. The accumulated dose mass was recorded by the microbalance during the dispensing. A staircase curve of accumulated dose mass against dispensing time is shown in Figure 2a. As the microbalance is stabilized during the interval of every two pulse signal, it gives a platform in the curve. The height between every two platforms gives the dose mass of dispensed samples. In essence, the discrete doses are uniform during the dispensing if the staircase is forwarded evenly and the gradient of the curve remains constant. The micro-dispensing was much quicker from 0.7 mm nozzle as indicated by an increase in dose mass. However, a decrease in accumulated dose mass was noted in the micro-dispensing of InhaLac®70 from 0.8 mm nozzle (Figure 3a). Subsequently, a significant increase in dispensed dose mass from 0.9 mm nozzle was observed with the staircase curve appearing linear. These changes were also reflected in the calculated mean dose masses as shown in Figure 2b. A dispensed mean dose mass range of 0.68 – 7.35 mg of InhaLac®70 could be achieved with the use of different nozzles. Mean powder flow rates of 6.8, 20.8, 16.0 and 73.5 mg/s at nozzle sizes 0.6-0.9 mm respectively were calculated by dividing mean dose dispensed through each nozzle by pulse vibration of 0.1s.

Attempts have been made to predict flow rate of powders from orifices of different shapes and sizes. Matsusaka and team studied the micro-feeding of fine powders using a vibratory capillary tube and found that the maximum flow rate of powders was dependant on the outlet capillary diameter or orifice and the kind of powders.[15] Beverloo and co-workers derived a general equation (Equation 1) to predict flow rate of free flow granules (1.2 mm – 1.6 mm) from circular orifices (2.6 mm - 30 mm) in a cylinder.[16] In the present study, InhaLac®70 as reported earlier[13] appeared as large agglomerated crystals (Figure 3b) having a smooth surface with low fine content. The particle size distribution was unimodal with average size (d_{50}) of 200 µm. With a reported bulk density of 0.59 g/cm^3 and present study involving dispensing of powders through a circular nozzle or orifice, flow rates at different nozzle or orifice sizes was estimated using equation 1.

$$W = 35 \, \rho_B \, \sqrt{g} \, (D_0 - 1.4d)^{2.5} \tag{1}$$

where W is the flow rate in g/min, ρ_B is the bulk density of powder in g/cm^3, g is the gravitational acceleration, D_0 is the diameter of the circular orifice, d is the particle size.

Flow rates of 0.12, 0.35, 0.26 and 1.04 mg/s using nozzle sizes of 0.6, 0.7, 0.8 and 0.9 mm could be estimated. As the estimated values were far smaller than real flow rate values, it meant that this equation did not fit the flow rate of small particles flowing from much smaller orifices used in this paper. A modified Equation 2 was proposed, this can fit the flow rate of InhaLac®70 from the nozzle sizes of 0.6 mm, 0.7 mm and 0.9 mm except nozzle size of 0.8 mm, which will require a further indepth study in the future.

$$W = 8.5 \; \rho_B \; \sqrt{g} \; (D_0 - 2.2d)^{2.3} \qquad (2)$$

Lu and co-workers in their studies with different powders have shown that acoustic vibration controlled micro-feeding depended on the balance between compaction and dilation of the powder in the nozzle.[6] Powders such as H13 tool steel powder and tungsten carbide showed stable micro-feeding in smaller nozzle sizes under acoustic controlled vibrations at a fixed voltage. The micro-feeding was more controlled and stable at 0.21 mm and 0.35 mm than at 1.35 mm. The authors speculated that arches of lower strength could form in larger nozzles leading to dispensing of irregular dose masses. In addition, stable micro-dispensing in acoustic energy controlled devices is dependent on behaviour of powder in the nozzle. The arching behaviour of powders in as columnar rods, discrete or cluster of particles has been found to be the most preferred one.

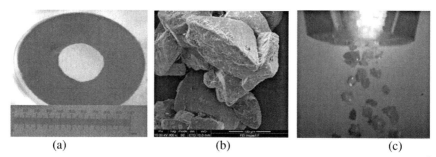

| (a) | (b) | (c) |

Figure 3   *(a) 0.8mm nozzle size; b) Scanning electron photomicrograph of InhaLac®70; (c) High speed image of InhaLac®70 falling as dispersed discrete particles from the 0.8mm nozzle*

The high speed photographic image (Figure 3c) captured during dispensing from a 0.8 mm nozzle size shows that InhaLac®70 particles fell as discrete particles. No distinct change in morphology could be noted. Further the flow ceased as the vibration was stopped. The preliminary results suggest that the micro-dispensing of InhaLac®70 particles was more stable in the nozzle sizes of 0.6 and 0.7 mm possibly due to formation of arches which could be broken easily to give uniform doses. The irregular dispensing pattern noted in the 0.8 mm nozzle could be due to residence of undischarged powder which steadily compacted on multiple pulses leading to lower dose masses. A further increase in the nozzle size (0.9 mm) may have caused formation of low strength arches leading to linear and significant increase in accumulated dose and mean dose mass. As vibration amplitude and oscillation duration have been identified as equally important parameters controlling microfeeding of powders, their effect on the dispensing of InhaLac®70 particles using our system will be independently studied in order to identify parameters providing the best microfeeding.[6]

4. CONCLUSIONS

The present study shows that use of different nozzles of an acoustic vibration controlled micro-dosing system with a 0.1 s pulse of vibration and a fixed voltage of 5 V could aid in obtaining a mean dose mass range of 0.68 – 7.35 mg of InhaLac®70. The discrete flow of InhaLac®70 particles suggested formation of arches at smaller nozzle sizes. The slow increase in dose mass under multiple pulses of vibration in larger nozzle sizes suggested irregular dose dispensing. The estimated flow rates increased on increasing nozzle sizes with non linear increase at > 0.7 mm. The inconsistent behaviour of particles at different nozzle sizes indicated that the selection of nozzle is critical for acoustic controlled micro-dispensing of micron-sized InhaLac®70. A new Beverloo format equation is proposed to predict the flow rate of InhaLac®70 from 0.6, 0.7 and 0.9 mm nozzles. The results of the study provide a scope for feasibility studies optimizing other important parameters such as vibration amplitude and duration of vibration of acoustic controlled micro-dosing system in micro-dispensing of inhalation grade micron-sized pharmaceutical materials.

References

1   Y. Jiang, S. Matsusaka, H. Masuda and Y. Qian, *Powder Technol.,* 2009, **188,** 242.
2   J.H. Bell, P.S. Hartley and J.S.G. Cox, *J. Pharm. Sci.,* 1971, **60,** 1559.
3   E.M. Littringer, A. Mescher, H. Schröttner, P. Walzel and N. A. Urbanetz, *Sci Pharm.,* 2010, **78,** 672.
4   X.M. Zeng, G. P. Martin, C. Marriott and J. Pritchard, *J. Pharm. Pharmacol.,* 2000, **52,** 1211.
5   S.F.Yang, *WO/2011/020862.,* 2011.
6   X. Lu, S.F. Yang and J. R. G. Evans, *J. Phys. D: Appl. Phys.,* 2006, **39,** 2444.
7   S.F. Yang and J.R.G. Evans, *Powder Technol.,* 2004, **142,** 219.
8   S.F. Yang and J.R.G. Evans, *Phil. Mag.,* 2005, **85,** 1089.
9   W.E. Deming and A.L. Mehring, *Industr. Engng. Chem.,* 1929, **21,** 661.
10 R.H. Newton, G.S. Dunham and T.P. Simpson, *Trans. Amer. Inst. Chem. Engng.,* 1945, **41,** 918.
11 R.L. Brown and P.G.W. Hawksley, *Fuel,* 1947 **26** 159.
12 X. Lu, S.F. Yang and J.R.G. Evans, *Powder Technol.,* 2007, **175,** 63.
13 G. Pilcer and K. Amighi, *Int. J.Pharm.,* 2010, **392,** 1.
14 N.D. Cekić, S.D. Savić, J. Milić, M.M. Savić, Ž. Jović and M. Maleševic, *Drug Del.,* 2007, **14,** 483.
15 S. Matsusaka, K. Yamamoto and H. Masuda, *Adv. Powder Technol.* 1996, **7,** 141.
16 W.A. Beverloo, H.A. Leniger, J. Van de velde, *Chem. Eng. Sci.,* 1961, **15,** 260.

THE INFLUENCE OF FLOW PROPERTIES ON ROLL COMPACTION BEHAVIOUR OF PHARMACEUTICAL BLENDS

N. H. Hamidi and C.-Y. Wu

School of Chemical Engineering, University of Birmingham, Birmingham, B15 2TT, UK.

1 INTRODUCTION

Granulation is a typical unit operation in the pharmaceutical industry to produce granules of large sizes from fine powder blends so that the density and flowability of the formulations can be increased and the process performance can be improved with a reduction in segregation tendency. There are two types of granulation processes: wet and dry granulation. In wet granulation liquid binders are generally used to agglomerate fine particles, while in dry granulation no liquid binder is needed and fine particles are brought together by compaction at high pressures. Roll compaction is an automated, continuous and efficient dry granulation process that has been widely utilized in pharmaceutical, chemical, foundry and mineral industries. It is also an effective process to increase the bulk density of formulations and to minimize the dust problem during the manufacture processes. Moreover, it is a preferable process for the formulations that are sensitive to heat or moisture because neither additional heat nor liquids are required in this process.[1,2]

During roll compaction, powder blends are fed into the gap between two counter-rotating rollers, which is sealed using either side cheek (sealing) plates or the rim on one of the rollers. Different feeding mechanisms can be used to feed a powder, such as gravity feeding, in which the powder is fed under gravity, and screw-feeding, in which the powder is fed with a rotating screw. The powder is gripped and squeezed between the rollers as a result of the shear forces generated by the friction between powders and the rotating rollers and the pressure induced by the screw feeder (or gravitational force). With different roll designs (fluted, knurled, smooth or pocket rolls), the powder is then compacted into ribbons, flakes or briquettes that are subsequently milled into granules of desirable sizes.

The properties of the produced granules (i.e. size distribution and strength) depend on a number of factors, including milling conditions and the ribbon properties (density distribution and strength). For a given milling condition, the porosity and size of the granules are determined by these properties of ribbons.[3] In general, smaller granules are produced from weaker ribbons that posses low mechanical strength while coarse ones from stronger ribbons under the same milling conditions.[4] There is a strong correlation between the strength of compressed pharmaceutical compacts (tablets and ribbons) and their relative densities.[5,6] Hence, the density distributions of roll compacted ribbons play an important role in controlling the granule properties. Ribbons with uniform density distribution are desirable in pharmaceutical manufacturing for controlling the quality of pharmaceutical granules. However, it is a very challenging task to produce such ribbons because the roll compaction behaviour is governed by the materials properties of formulations and the roll compaction conditions.

It is often observed in pharmaceutical manufacturing process that when a formulation with good flowability is processed using a roll compactor, the powder can pass through the roll gap easily so that the powder is not well compressed to ribbons that are dense enough for producing granules of desirable sizes. On the other hand, if a formulation is cohesive and has a poor flowability, it is very difficult to feed the powder into the compaction zone between the rollers. Consequently, only a limited amount of powder is gripped into the compaction region and the produced ribbons are either very loose (i.e., with low densities) when a fixed roll gap is used, or very thin when floating rollers are used. The flow properties of pharmaceutical formulations can play an important role in roll compaction process. However, the correlation of flow properties with the roll compaction behaviour is not clearly understood. Thus the aim of this study is to explore how the flow properties and frictional behaviour of powder blends affect the roll compaction behaviour and subsequently the properties of roll compacted ribbons.

2 MATERIALS AND METHODS

Two common pharmaceutical excipients, microcrystalline cellulose (MCC, Avicel PH-102) and lactose monohydrate (LMH), were chosen as model materials. MCC is the most widely used excipient that serves a number of functions in tablets, such as diluents, disintegrant, sorbent and anti-adherent, while lactose monohydrate works as a diluents and flow improvement agents.[7] When being compressed, MCC particles deform plastically and are held together by hydrogen bonds,[8] which makes MCC compacts stronger and easy to be compacted directly, while lactose particles show a strong tendency to fragment during compaction. In addition, the bonding strength of lactose particles is much lower than MCC particles,[5,6] even though lactose monohydrate has a better flowability.

The as-received MCC and lactose were mixed at different concentrations in order to systematically vary the flow properties of powder mixtures, which were measured using a ring shear cell tester (RST-XS Dr Dietmar Schulze Wolfenbuttel Germany). Shear cell tests were performed with a preload normal stress of 6 kPa, 8 kPa and 10 kPa for each sample, and the flow function and effective internal frictional angle are then determined. The powder mixtures were then roll compacted using the instrumented roll compactor developed at the University of Birmingham,[1,2] in which the two rollers are sealed using two side cheek plates. For the roll compaction tests, the roll gap was set at 1.0 mm. The powder mixtures were feed into the compaction zone through a feeder hopper as the two rollers were kept stationary. It was interesting to notice that some powder mixtures (typically with higher concentrations of LMH) flowed through the roll gap freely before the roll compaction was started and consequently cannot be compacted. For those powder mixtures that can completely fill the compaction zone and the feeder hopper, the powders were compacted at a roll speed of 1 rpm. The powder flow behaviour during roll compaction was recorded using a video camera. The density distributions of produced ribbons were determined using the sectioning methods.[2] The correlations between the flow properties of powders, the powder flow behaviour during roll compaction and the density distributions of ribbons were examined.

3 RESULTS AND DISCUSSION

3.1 Powder Flowability

The powder flowability is characterized in terms of flow function that is defined as the ratio of consolidation stress to the unconfined yield stress. For various powder mixtures considered in this study, the flow functions were determined using the ring shear cell tester, which was also used to measure the internal frictional angles. Figure 1 shows the variations of flow function and the internal frictional angle with the concentration of MCC in the powder mixtures. It is clear that the flow function decreases while the internal frictional angle slightly increases, as the concentration of MCC increases. This indicates that the internal friction increases and the flowability of the powder mixtures decreases with the increasing concentration of MCC. However, all powder mixtures have good flowability and can be classified as e either free flowing (flow function is larger than 10) or easy flowing (flow function is between 4 and 10). Specifically, when the concentration of MCC is higher than 70%, the powder mixture is easy flowing.

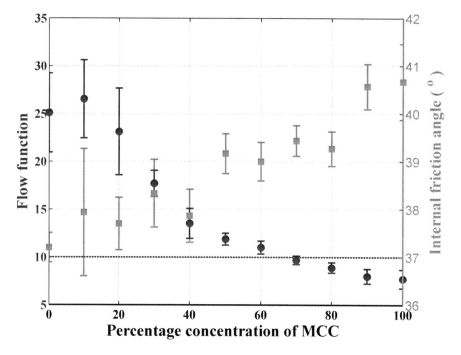

Figure 1   *The flow and frictional properties of the powder blends with various concentrations of MCC (Circle: flow function; Square: internal frictional angle). The error bars show the standard deviation of three measurements. (Circle – flow function; Square – Internal frictional angle)*

3.2 Roll Compaction Behaviour

When the powder mixtures were fed into the feed hopper of the roll compactor while the rolls were stationary, it was observed that the powder mixtures of low and intermediate concentration of MCC (say less than 60%) flowed through the roll gap (of 1mm) freely (see Figure 2a & 2b). While the mixtures with high concentration of MCC (70% or higher) stayed in the feed hopper so that they were roll compacted once the hopper was full and the rolls started to rotate. This is attributed to the powder mixtures with different flowability and frictional properties as the concentration of MCC is varied, as shown in Figure 1. For the case considered in this study, it is found that free flowing powder mixtures (flow functions >10) can flow freely through the roll gap and cannot be roll compacted with the approach adopted, while easy flowing powder mixtures (flow function < 10) did not flow through the roll gap so that they can be compacted to form ribbons.

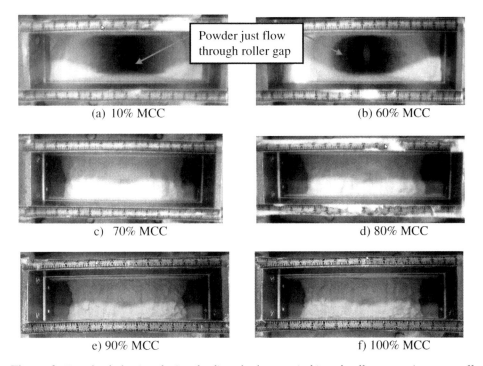

(a) 10% MCC   Powder just flow through roller gap   (b) 60% MCC

c) 70% MCC   d) 80% MCC

e) 90% MCC   f) 100% MCC

Figure 2 *Powder behavior during feeding the hopper (a,b) and roll compaction at a roll speed of 1 rpm (c-f) for powder mixtures with different concentration of MCC.*

The flow behaviour of these powder mixtures in the feed hopper during the roll compaction (i.e., when the two roller are counter-rotating) was recorded. The powder profiles at the same degree of fill of the powder in the roll gap are presented in Figs. 2c-2f. It is clear that there is a dip all along the powder surface at the middle of the roll width so that "v-shaped" patterns are obtained when the powder profiles are projected onto the roll surface.[1,2] This is due to the influence of friction between the powder and the side cheek plates, which inhibits the flow of the powder in the vicinity of the side cheek plates. Consequently, the powder close to the middle of roll width is dragged into the compaction zone at a higher rate than those at the edges. This demonstrates that the powder does not flow uniformly during the roll compaction process. The direct consequence of the non-uniform flow of powder during roll compaction is the heterogeneity of ribbon density distribution. A high flow rate of powder in the middle of roll width implies that the powder is compressed with a higher powder mass flux in this region. Consequently, the ribbon density in the middle is higher than that at the edges, as shown in Figure 3. This non-uniform density distribution is believed to be responsible for the wide granule size distribution obtained during milling. The low density region at the edges of the ribbon would result in an excessive concentration of fine granules.

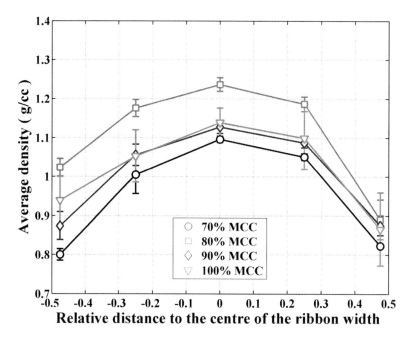

Figure 3   *Density variation along the ribbon with for the powder blends with various concentrations of MCC (roll speed 1rpm)*

4 CONCLUSIONS

The flowability and frictional properties of binary mixtures of microcrystalline cellulose and lactose were varied systematically by altering the concentration of MCC in the mixture, and were measured using a ring shear tester. The produced powder mixtures were then fed into a laboratory-scale instrumented roll compactor, the flow behaviour was recorded and the density distributions of produced ribbons were also characterized. It has been found that, for the cases considered in this study, there is a good correlation between the process-ability (i.e., flow/no-flow situations) during the roll compaction and the flowability measured from shear testing. Furthermore, it has been found that how the powder is fed into the compaction zone determines the ribbon density distribution. The friction between the powder and the fixed side cheek plates inhibits the flow of powders, hence the powder in the middle of roll width flow faster and are compressed into a much denser state when compared to those at the edges.

References

1. A.M. Miguelez-Moran, C.-Y. Wu and J.P.K. Seville, *Int.. J. Pharm.*, 2008, **362**, 52.
2. A.M. Miguelez-Moran, C.-Y. Wu, H. Dong and J.P.K. Seville, *Euro. J. Pharm. Biopharm.*, 2009, **72**, 173.
3. C. Nystrom and G. Alderborn, *Drug. Dev. Ind. Pharm.*, 1993, **19** (17-18), 2143.
4. S. Inghelbrecht and J.P. Remon, *Int. J. Pharm.*, 1998, **171**, 195.
5. C.-Y. Wu, S.M. Best, A.C. Bentham, B.C. Hancock and W. Bonfield, *Eur. J. Pharm. Sci.* 2005, **25**, 331.
6. C.-Y. Wu, S.M. Best, A.C. Bentham, B.C. Hancock and W. Bonfield, *Pharm. Res.* 2006, **23**(8), 1898.
7. C.-Y. Wu and J.P.K. Seville, *Powder Technology*, 2009, **189**, 285.
8. H.A. Lieberman, L. Lachman and J.B. Schwartz, *Pharmaceutical dosage forms: Tablets.* Informa Health Care: 1990.

AN EXPERIMENTAL INVESTIGATION OF DRY GRANULATION

C.H. Goey, K. Lu and C.-Y. Wu

School of Chemical Engineering, University of Birmingham, Edgbaston, Birmingham B15 2TT, UK

1 INTRODUCTION

Dry granulation is a widely utilized agglomeration process in pharmaceutical, chemical, foundry and mineral industries, which generally consists of roll compaction that compresses powder blends to ribbons or flakes, and milling that break ribbons/flakes into granules (Figure 1). It is an automated, continuous and efficient process that produces granules with improved flowability and homogeneity,[1,2] which are important in facilitating the downstream manufacturing process.[3] In addition, dry granulation is effective in minimizing dust problem by increasing the bulk density and particle size of powders.

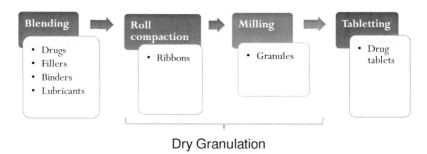

Figure 1 *Typical pharmaceutical manufacture process involving dry granulation*

The ribbon and granule properties are influenced by the process conditions (i.e. roll speed, roll gap, feeding mechanism, feeding speed, and milling rate) and the powder properties (i.e. particle size, shape and moisture content). On the other hand, ribbons are inevitably exposed to attrition during milling process and this will in turn affect the granule properties. However, the impact of attrition on the properties of ribbons and granules are still not known. In this study, the effect of powder particle size on ribbons properties (i.e. solid fraction and density distribution), granular properties (i.e. granular size distribution) and ribbon attrition were investigated in detail.

2 MATERIALS AND METHODS

2.1 Materials

Two widely used pharmaceutical excipients, microcrystalline cellulose of Avicel grade PH-101 (MCC101) and PH-102 (MCC102) (FMC Biopolymer, USA), are used as model materials. MCC101 powder has a smaller particle size than MCC102 powder. Table 1 presents the general properties of MCC101 and MCC102.

Table 1 *Particle Size and bulk density of MCC101 and MCC102*

Powder type	Mean Diameter (µm)	Bulk Density (x10^3 kg/m^3)
MCC101	50	0.29
MCC102	90	0.31

2.2 Roll Compaction and Ribbon Characterization

Roll compaction was performed using a purpose-built gravity-fed roll compactor developed at the University of Birmingham. The dimensions of the rolls were 45 mm wide and 200 mm in diameter. A detail description of this roll compactor was provided in previous publication.[6] The roll speed was set at 3 rpm for all the experiments reported here.

Roll compaction was carried out in a fixed initial condition. Before the roll compaction, a constant volume of powder was transferred into a hopper until a small heap was formed over the top of the hopper and then levelled off. Thereafter, the roll compaction was initiated where the powder was gravity fed into the compaction zone by two counter-rotating rollers.[2] As suggested by previous authors, the ribbons produced at the beginning and the end of the roll compaction were discarded and only those in between were chosen for further analysis.[7] During roll compaction, a digital camera (Olympus FE-204) was employed to record the powder flow patterns in the hopper.

Evolution of the compression pressure was recorded with pressure transducer that was fitted onto the roll surface. The compression pressure was collected every 0.39° of rotation. The data were processed using MATLAB to obtain the pressure profiles, from which the nip angle and the maximum compression pressure were determined using the same approach as reported by previous authors.[4,5,6,7]

Density distributions for the ribbons produced with MCC101 and MCC102 were also determined using a sectioning method. Ribbon strips with approximate dimensions of 6 mm x 43 mm (Figure 2) were obtained using a bandsaw (Scheppach Basato 1, Germany) and weighed using a balance (Ohaus Explorer Precision EP213, USA). The exact dimensions of the ribbon (length x width x thickness) were measured using a slide calliper (Mitutoyo CD-6" CP). Before sectioning, one end of the ribbon strip was marked to indicate the sectioning direction. A small section (ca. 5 mm wide, see Figure2) was then sawn off from the ribbon strip (e.g. Section "1" in Figure 2). The remaining ribbon (i.e. Section '2' to '7' in Figure 2) was weighed and its width was measured. The sectioning and measuring procedures were repeated so that the mass and dimensions for each sectioned piece were determined. The bulk density of each section was then calculated from the mass and the volume. The relative density was calculated as the ratio of the bulk density to the true density of the powder.[7]

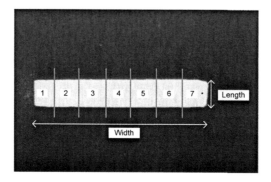

Figure 2 *Illustration of the sectioning method to determine the ribbon density distribution*

2.3 Milling and Granule Characterization

The produced ribbons were milled using a cutting mill (Retsch SM 100, Germany) equipped with a 4 mm square sieve at a speed of 1500 rpm. The produced granules were collected for further analysis.

The granular size distribution was evaluated with sieve analysis. For each type of granules, the granule size distribution was carried out by sieving 100 g of granules on the sieve shaker for 10 minutes at amplitude of 1 mm, with stacked sieves of 50, 63, 100, 180, 280, 630, 1000, 1400 and 2000 μm (Retsch, Germany). The mass of granules remaining on each sieve was weighed and the corresponding size distribution was determined.

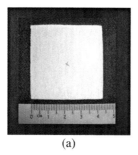
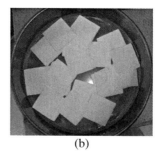

(a)   (b)

Figure 3 *Ribbons prepared for attrition tests*

2.4 Attrition Tests

Ribbon attrition tests were carried out at an attrition amplitudes 3 mm. During the test, 50 g of ribbons was cut into squares of 43 mm x 43 mm (see Figure 3a) with the bandsaw. These ribbons were placed onto a sieve with an aperture of 2000 μm and shaken on a sieve shaker (Retsch AS200, Germany), with a 5 minute interval up to 25 minutes. Ribbons remained in the sieve were weighed at each time interval using the balance, from which the attrition behaviour (mass loss and attrition rate) was examined.

3 RESULTS AND DISCUSSION

3.1 Roll Compaction

The powder flow patterns during roll compaction of MCC102 and MCC101 at roll gap of 0.8 mm are shown in Figure 4 and Figure 5, respectively. For both cases, the initial powder bed had a flat surface. During roll compaction, V-shaped patterns were observed over the surface of the MCC102 powder bed (Figure 4), which is consistent with the observations reported in previous publication.[7] This indicates that the powder located at the edges of the rollers is less dense due to friction with the side cheek plates. Consequently, as confirmed by other researchers, ribbons produced by roll compaction with stationary cheek side plates generally have lower densities at the edges than in the middle.[7,8]

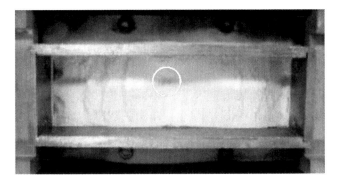

Figure 4 *Formation of a dip in the powder bed, leading to a V-shaped profile during roll compaction of MCC102 at 0.8 mm roll gap*

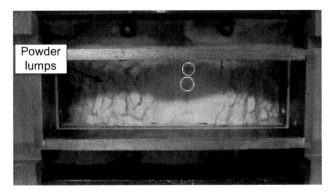

Figure 5 *Formation of the two dips in the powder bed, leading to a 'W-shaped' profile during roll compaction of MCC101 ribbons at 0.8 mm roll gap*

On the other hand, an interesting phenomenon was observed during the roll compaction of MCC101 (Figure 5), where a W-shaped profile was observed instead of a V-shaped profile. It shows that there were two dips along the powder surface across the roll width, where the powder was drawn into the compaction zone faster than the powder located at the middle and the edges of the rollers. Moreover, the powder was observed to be flowing in lumps, cascading down into the compaction zone.

The difference in flow profiles of MCC101 and MCC102 is believed to be caused by the difference in particle size. Since MCC101 has a smaller particle size compared to MCC102, the interparticulate forces in MCC101 are stronger and the particles will tend to bond together and flow intermittently. This complicates the powder flow in the sense that it is harder for the particles to be gripped by the rolls.

3.2 Density Distribution and Solid Fraction of Ribbons

Figure 6 presents the density distributions of ribbons made of MCC101 and MCC102. In general, the ribbon produced by roll compaction has a non-uniform density distribution across its width, with lower densities at the edges and higher densities in the middle, which are produced due to inhomogeneity in the flow profiles as described in Section 3.1. Large standard deviations in the density measured are due to two possible reasons: (1) the accuracy and consistency are limited by the experimental method employed and (2) the degree of variation in localized density across the ribbon width is in fact very high.

It is interesting to note the prominent differences between the density distributions between MCC101 and MCC102 ribbons. Density distributions of MCC101 are generally higher than MCC102. Moreover, two peaks are observed in the density distributions of MCC101 while only a single peak is noticed in the density distribution of MCC102, which is consistent with the flow profile described in Section 3.1, implying that powder flow patterns during roll compaction determine the density distributions of ribbons.

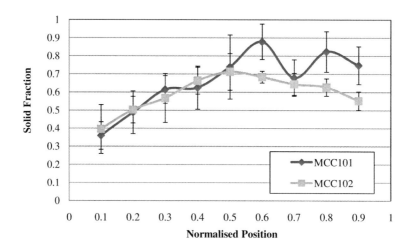

Figure 6 *Density distributions of ribbons of MCC101 and MCC102 at 0.8 mm roll gap*

Figure 7 shows the mean solid fractions (i.e. relative density) for all types of ribbons considered in this study. It is clear that MCC101 ribbons generally have higher solid fraction than MCC102. This is because MCC101 has a smaller particle size that increases the interparticulate contact and bonding area, and hence form a denser packing and consolidate state. Consequently, a higher ribbon solid fraction is obtained for MCC101 compared to MCC102. In addition, the solid fraction decreases as the roll gap increases, as observed by many others.[11]

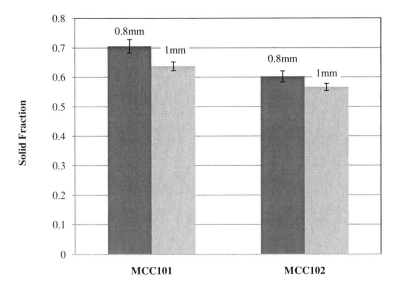

Figure 7 *Average solid fractions for ribbons of MCC101 and MCC102 produced at 0.8 mm and 1 mm roll gaps*

3.3 Mean Granular Size and Span of Granular Size Distribution (GSD)

Figure 8 presents the mass mean diameter of granules of MCC101 and MCC102 produced at 0.8 mm and 1 mm roll gaps. It can be seen that mean granular size of MCC101 is slightly higher than MCC102, although the primary MCC101 particles are smaller than MCC102. In addition, as the roll gap increases, the mass mean diameter reduces. This is because under the similar milling condition, the granular size is strongly dependent upon the relative density of the ribbons, where ribbons with higher solid fractions generally produce larger granules.

Figure 9 shows the span of the granule size distributions for all four types of ribbons examined. A higher value of span indicates a wide particle size distribution. It is clear that smaller spans were obtained for the granules produced from the ribbons roll-compacted with a narrower roll gap (say 0.8 mm). This is attributed to the higher ribbon solid fractions when the powder is compacted at a narrower roll gap (see Figure 7). Nevertheless, it is noticed that, for the granule produced using the ribbon compressed at 0.8 mm roll gap, the span for MCC101 granules is slightly higher than that of MCC102, even though the corresponding MCC101 ribbon has a higher solid fraction than the MCC102 ribbons (see

Figure 7). A close examination of the density distribution (see Figure 6) reveals that the degree of variation in the localized density of MCC101 ribbon (ranges from 0.35 to nearly 0.9) is much greater than MCC102 (ranges from 0.4 to a maximum value of 0.72). It is believed that this larger variation in the localized density of MCC101 ribbon is responsible for the larger span of MCC101 granules compared to MCC102.

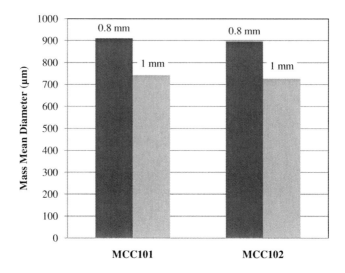

Figure 8 *Mass mean diameters for various MCC101 and MCC102 granules considered*

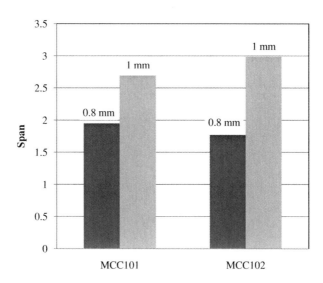

Figure 9 *Corresponding spans for granule size distributions of various MCC101 and MCC102 granules*

3.4 Attrition

The typical evolution of ribbon shape during attrition tests is illustrated in Figure 10 and the corresponding instantaneous mass loss is shown in Figure 11. At the early stage of attrition tests, large amount of fines which were loosely attached to the edges and surfaces of ribbons are attrited and this contribute to high ribbon mass loss (Figure 11) even though there is no apparent change in the ribbon shape (Figure 10a to 10b). As the attrition process progressed, small fragments were chipped off from the ribbon instead of more fines, resulting in dramatic change in ribbon shape from square to oblong (Figure 10e to 10f), during which ribbon breakage becomes dominant. To some extent, the change in ribbon shape during attrition implies the inhomogeneous density distribution across the ribbon width. It can also be seen from Figure 11 that the larger instantaneous mass loss is obtained for MCC101 than MCC102. This is primarily due to the difference in ribbon solid fraction as shown in Figure 6. This indicates that the ribbon solid fraction dominates the milling and attrition processes.

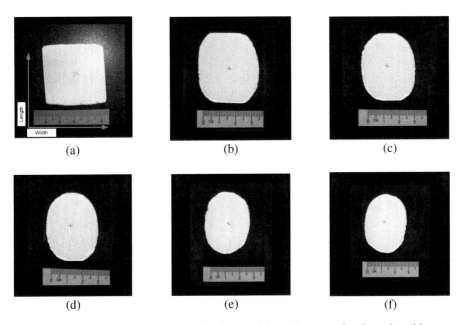

Figure 10 The change in ribbon shape during *attrition. Photographs show the ribbon shape at 5 minute intervals, from (a) 0 min to (f) 25 minutes*

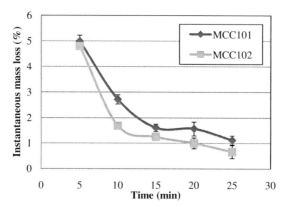

Figure 11 *Instantaneous ribbon mass loss at 5 minute interval for MCC101 and MCC102 ribbons produced at roll gap of 1 mm*

4 CONCLUSIONS

The effect of powder particle size on dry granulation of MCC powders was investigated. It has been found that the powder with smaller particle size tends to flow intermittently, in which the powder cascades onto the roll surfaces as chunks of agglomerates.

For MCC101, the stagnant zone created on the powder bed during roll compaction resulted in a W-shaped flow profile, and consequently a density distribution with two peaks instead of one peak, as normally observed in MCC102. This indicates that the flow pattern during roll compaction has a significant effect on ribbon and granule properties. It has also been found that powder of smaller particle size will be compacted into ribbons with higher solid fraction. These properties of ribbon in turn affect the granular properties in the sense that stronger ribbons will produce coarser granules with a smaller span in granule size distribution (i.e. narrower granule size distribution). In addition, it is observed that greater variations in the local density of ribbons lead to a wider granular size distribution.

It has also been shown that a large amount of mass loss occurred at the initial stage of attrition, where erosion is dominating the attrition process. At the later stage, mass loss was mostly due to small fragments chipping off from the ribbon edges where ribbon breakage becomes dominant.

References

1   F. Freitag and P. Kleinebudde, *Eur. J. Pharm. Sci.*, 2003, **19**, 281.
2   C.-Y. Wu, W. L. Hung and A.M. Miguelez-Moran, *Int. J. Pharm.*, 2010, **391**, 90.
3   C. Bacher, P.M. Olsen and P. Bertelsen, *Int. J. Pharm.*, 2007, **342**, 115.
4   S.-J. Wu, C. Sun, *J. Pharm. Sci.*, 2007, **96**, 1445.
5   R.F. Mansa, R.H. Bridson, R.W. Greenwood, *Powder Technol.*, 2008, **181**, 217.
6   G. Bindhumadhavan, J.P.K. Seville, M.J. Adams, *Chem. Eng. Sci.*, 2005, **60**, 3891.
7   A.M. Miguelez-Moran, C.-Y. Wu, and J.P.K. Seville, *Int. J. Pharm.*, 2008, **363**, 52.
8   Y. Funakoshi, T. Asogawa and E. Satake, *Drug Dev. Ind. Pharm.*, 1977, **6**, 555.
9   S. Malkowska and K.A. Khan, *Drug Dev. Ind. Pharm.*, 1983, **9**, 331.
10 P.J. Sheskey and J. Hendren, *Pharm. Technol*, 1999, **23**, 90.

11 A.M. Miguelez-Moran, C.-Y. Wu, H. Dong and J.P.K. Seville, *Euro. J. Pharm. Biopharm.*, 2009, **72**: 173.

EXPERIMENTAL INVESTIGATION OF MILLING OF ROLL COMPACTED RIBBONS

S. Yu,[1] B. Gururajan,[2] G. Reynolds,[2] R. Roberts,[2] M. J. Adams[1] and C.-Y.Wu[1]

[1] School of Chemical Engineering, University of Birmingham, Edgbaston, Birmingham, B15 2TT
[2] Pharmaceutical Development, AstraZeneca, Macclesfield, Cheshire, SK10 2NA

1 INTRODUCTION

Roll compaction is widely adopted in pharmaceutical industries for dry granulation. The roll compacted ribbons are milled to form granules in order to meet the requirements for subsequent tabletting or capsulation. In practice, a mill is commonly integrated in the roll compactor to produce granules. [1, 2] The size distribution of the granules is an important characteristic since it affects the mechanical properties and uniformity of the final product. Moreover, in the design of a continuous process, the efficiency of the milling process is critical for achieving throughput targets. Therefore, it is important to understand the underlying mechanisms in the milling process, and the factors that govern the size distribution and mass throughput of the granules. With a better understanding of the process, there is a potential to predict the roll compaction and milling parameters required to achieve the minimum residence time in the milling chamber, the maximum throughput, and the specified granule size distribution (74 – 840 μm). [3]

An oscillating granulator is an instrument for mechanically passing compacted materials (roll compacted ribbons) through a wire mesh screen using an oscillating rotor [3]. The size of the milled granules is controlled by the screen size, rotor speed, and rotational angle of the rotors.[4] Abraham and Cunningham[5] reported that an oscillating granulator could be used as a downstream granulator to provide a formulation with improved flow characteristics and a reduced amount of fine particles. They showed that such equipment produced coarser granules comparing to a comil or flitzmil with the same mesh size. Sakwanichol et al.[6] evaluated the granule properties obtained from an oscillating granulator and a two-stage roll mill. They examined four oscillating granulator parameters i.e. rotor speed, oscillating angle, aperture of mesh screen and rotor type, and six roll-mill parameters, i.e. throughput, speed ratio in both first and second stages, gap between roll pair in both stages and roll-surface texture. They compared the granules with similar median particle size produced with both mills. They showed that that, for the oscillating mill, all the examined parameters affected significantly the granule size distribution. However, the rotor type of the oscillating granulator did not affect the amount of fines. Fonner et al.[7, 8] examined the pharmaceutical solids prepared by 5 different granulation methods and reported that the flowability, as characterised by the angle of repose, was primarily a function of the surface roughness of the granules. All previous studies illustrated the importance and complexity of milling processes.

It is important to understand the mechanisms in the milling process that can dictate the granule and tablet properties. For this purpose, the emphasis of the current work is on the milling process using an oscillating granulator. The effects of the solid fraction and strength of the ribbons and the rotational speed of the blades on the milling performances were investigated.

2 MATERIALS AND METHODS

Microcrystal cellulose (MCC) of Avicel grade (PH102, FMC Biopolymer, USA), a crystalline powder (crystallinity > 78%) with needle-shaped particles, was selected as a model powder. The true density was measured using a helium pycnometer (AccuPycII 1340, Micromeritics USA).

The MCC was compacted using a laboratory scale instrumented roll compactor developed at the University of Birmingham.[9-11] The roll speed was set to 1 rpm and the minimum gap was varied in the range 0.8 - 1.4 mm in order to produce ribbons with different solid fractions. Ribbons of ~100 g were cut to specific dimensions (approximately 22 x 22 mm) in order to minimise the effects of variations in ribbon shapes and sizes. The dimensions (i.e. length, width and thickness) of such segments of the ribbons were measured accurately using a digital calliper (Mitutoyo, Hampshire, UK) to determine their volumes, from which the bulk density could be calculated from their mass. The solid fractions were obtained as the ratio of the bulk density to the true density. The fracture energies of the ribbons were measured by a 3-point bending tests using a universal mechanical testing machine (Instron, High Wycombe, UK). This involved integrating the force-displacement data to determine the total work to fracture. The fracture energy was obtained as the ratio of the work and the area of the fracture surface.

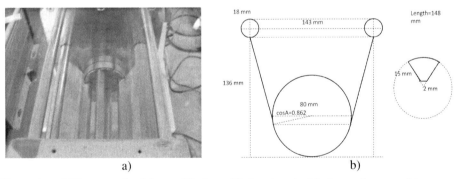

Figure 1 *a) Photograph of the oscillating mill showing the blades and screen; b) schematic diagram of the cross-section of the oscillating mill.*

An oscillating mill (Coeply, AR 401) was used for the milling (Figure 1) with a screen aperture size of 630 µm, and at milling speeds of 50 to 330 rpm. The volume of the milling chamber was maintained at a constant value. In order to prevent granules escaping from the sides of the screens, which caused inaccuracies in the mass throughput and an artifactually increase in the mean granule size, the sides of the screen were sealed with a plastic film. The mass throughput of granules as a function of time was measured using a computerised balance. The size distribution was measured using an image analysis method (QICPIC model, SympaTec).

3 RESULTS

The fracture energies of the ribbons as a function of the solid fraction are shown in Figure 2. The values increase exponentially according to the following relationship:

$$G_f = a\exp(b\phi) \qquad (1)$$

where a and b are constants. The line in Figure 2 is the best fit to the data with values of the constants $a = 1.31 \frac{J}{m^2}$ and $b = 6.76$. This relationship is consistent with previous studies of the strength of granules.[12]

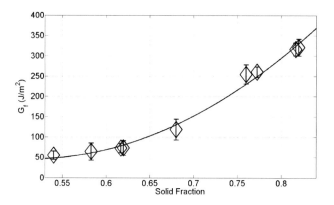

Figure 2   *Fracture energy of the ribbons as a function of the solid fraction. The line is the best fit to Eq. (2).*

Figure 3 shows typical data for the mass throughput as a function of the number of milling cycles. It is possible to describe the data using first order kinetics:

$$\frac{dm}{dN} = k\,(m_\infty - m) \qquad (2)$$

where N is the number of cycles, and m and m_∞ are the mass of granules for $N = N$ and $N = \infty$ respectively. Integrating subject $m = 0$ at $N = 0$ yields:

$$m = m_\infty\left[1 - \exp\left(-\frac{N}{N_c}\right)\right] \qquad (3)$$

where N_c is the characteristic number of milling cycles at which $m = m_\infty\left(1 - e^{-1}\right)$ and thus corresponds to the milling process being 63% complete. Consequently, smaller values of this parameter correspond to more rapid breakdown of the ribbons. The best fit to Eq. (3) of the data in Figure 3 is also given in the figure.

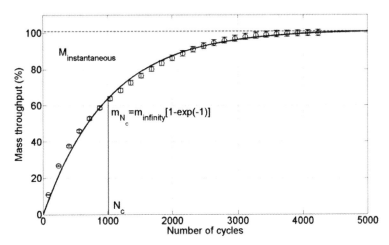

Figure 3 *Typical mass throughput profile for milling of roll compacted ribbons.*

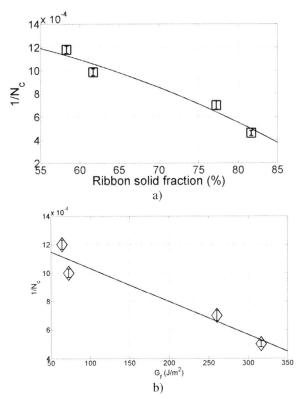

Figure 4 *The reciprocal of the characteristic number of milling cycles as a function of (a) ribbon solid fraction and (b) fracture energy of the ribbons.*

The value of N_c increases with increasing solid fraction and hence with the fracture energy of the ribbons as shown in Figure 4. That the number of milling cycles increases with the strength of the ribbons is clearly the expected result although it is not a simple linear relationship. It may be seen from the data in Figure 4a that an increase in the fracture energy by a factor of ~ 6 results in an increases in the number of cycles from ~ 800 to 2000, which is only a factor of 2.5.

Figure 5 shows that the mean particle size of the granules increases with increasing solid fraction (Figure 5a) of the ribbons and also the fracture energy (Figure 5b). Thus the milling time increases with the strength of the ribbons and weaker ribbons tend to fragment more, which results in a greater proportion of granules that are smaller than the aperture size of the mesh.

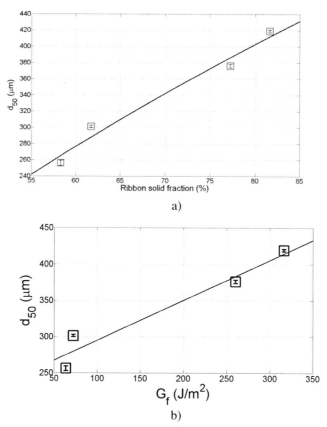

a)

b)

Figure 5 *The mean particle size of the granules as a function of (a) ribbon solid fraction and (b) fracture energy of the ribbons.*

Figure 6 shows data for $1/N_c$ as a function of the milling speed for ribbons having a solid fraction of 0.76. It appears that at small milling speeds the characteristic number of milling cycles is relatively independent of the speed and at greater speeds this number increases more rapidly.

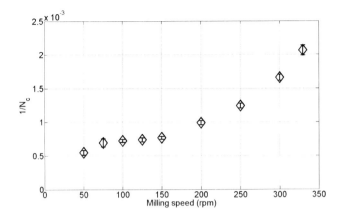

Figure 6   *The reciprocal of the characteristic number of milling cycles as a function of milling speed for the ribbons have a solid fraction of 0.76.*

Figure 7 shows data for granule size as a function of the milling speed for ribbons with a solid fraction of 0.76. It can be seen that the mean granule size decreases as the milling speed increases. Moreover, the mean sizes are less than the aperture value of the sieve (630 μm). At the highest speed, the mean size is ~ 2 less than the aperture size, which suggests that considerable fragmentation as occurred under these intense milling conditions.

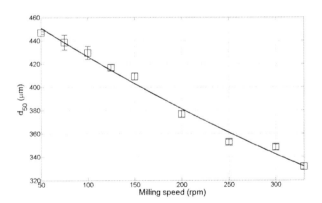

Figure 7   *The mean granule size as a function of milling speed for ribbons having a solid fraction of 0.76*

It was observed from the variation of $1/N_c$ with the milling speed and there appears to be two regimes. If it is assumed that the granulation rate is a function of the kinetic energy imposed by the rotors on the residual material in the mill due to an impact mechanism, it might be expected that a relationship of the following form would be applicable:

$$1/N_c = k\,m\,v^2 \tag{4}$$

where v is a characteristic speed of the rotors, k is a system constant and m is a function of the mass of the residual material (whole and damaged ribbons). The interpretation of this relationship is not trivial since the mass is a decreasing function of the number of cycles and it is not clear whether, say, the maximum or mean velocity of the rotators is the critical value. The data has been replotted in Figure 8 assuming that the mass is a constant that is equal to the initial value although clearly is likely to be a more complex function. Given that Eq. (4) is a close approximation in the high velocity regime, it appears that an impact mechanism is a reasonable interpretation. In the low velocity regime, the rate is much less sensitive to the velocity and it is arguable that it could correspond to an abrasion mechanism that is more predominantly a function of the distance over which the material has been displaced.

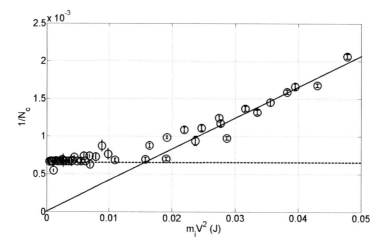

Figure 8   *The recipricol of the characteristic number of milling cycles as a function of kinetic energy where the full line represents the impact regime and the dased line represents the abrasion regime*

4 DISCUSSION

The strength of compacts is primarily a function of their porosity or the solid fraction.[13] Therefore, the solid fraction determines the strength of the roll compacted ribbons for a given formulation. Thus, the rate at which the ribbons are milled, as characterised by the parameter N_c, has a direct correlation with the solid fraction (see Figure 4a), as expected. When the milling speed is fixed, the energy input remains at a constant value if the thermal consumption is ignored, the fracture energy determines the breakage of the ribbons and the corresponding granule size is a linear function of the strength of the compacts. Due to the correlation between strength and solid fraction, the mean granule size is also determined by the solid fraction.

The initial mass of the feed ribbons was maintained at a constant value in the present work, so that any variation of the kinetic energy arises from changes in the milling speed. It was found that there are two distinctive regimes: (i) a quasi-static regime, where N_c is approximately constant when the milling kinetic energy is smaller than a critical value of the kinetic energy, E_c (~ 0.016 J for the case considered here), and (ii) an impact regime, in which $1/N_c$ increases linearly at kinetic energies greater than E_c. In the quasi-static regime, the ribbons move at a relative low speed since they are driven by the moving blade. In this case, abrasion is the dominant milling mechanism. In the impact regime, collisions between ribbons and between ribbons and the internal surfaces of milling chamber (*i.e.* blade, wire mesh) become dominant.

5 CONCLUSIONS

The mill performance of MCC ribbons was investigated experimentally. The effects of ribbon solid fraction and milling speed were examined. A first order kinetics equation was introduced to describe the relationship between mass throughput and the number of milling cycles. The critical number of cycles was defined to characterise the milling process. It was found that the mean granule size increases linearly with ribbon solid fraction but decreases linearly with the milling speed. Two distinctive milling regimes were identified: (1) quasi-static regime and (2) impact regime.

Acknowledgement

The authors would like to acknowledge the financial support from AstraZeneca.

References

1. M. G. Herting and P. Kleinebudde, *International Journal of Pharmaceutics*, 2007, **338**, 110-118.
2. P. Kleinebudde, *European Journal of Pharmaceutics and Biopharmaceutics*, 2004, **58**, 317-326.
3. T. A. Vendola and B. C. Hancock, *Pharmaceutical Technology*, 2008, **32**, 72-86.
4. R. J. Lantz, in *Pharmaceutical Dosage Forms: Tablets*, eds. H. A. Lieberman and J. B. Schwartz, Marcel Dekker, New York, Editon edn., 1990, vol. 2, pp. 107-200.
5. *US Pat.*, 2008.
6. J. Sakwanichol, S. Puttipipatkhachorn, G. Ingenerf and P. Kleinebudde, in *Pharm Dev Technol*, Editon edn., 2010.
7. D. E. Fonner, G. S. Banker and J. Swarbrick, *Journal of Pharmaceutical Sciences*, 1966, **55**, 181-186.
8. J. Sakwanichol, S. Puttipipatkhachorn, G. Ingenerf and P. Kleinebudde, *Pharmaceutical Development and Technology*, **0**, 1-10.
9. G. Bindhumadhavan, J. P. K. Seville, M. J. Adams, R. W. Greenwood and S. Fitzpatrick, *Chemical Engineering Science*, 2005, **60**, 3891-3897.
10. A. M. Miguélez-Morán, C. Y. Wu and J. P. K. Seville, *International Journal of Pharmaceutics*, 2008, **362**, 52-59.
11. B. A. Patel, M. J. Adams, N. Turnbull, A. C. Bentham and C. Y. Wu, 2010.

12. L. J. Vandeperre, J. Wang and W. J. Clegg 釰, *Philosophical Magazine*, 2004, **84**, 3689-3704.
13. C.-Y. Wu, S. M. Best, A. C. Bentham, B. C. Hancock and W. Bonfield, *European Journal of Pharmaceutical Sciences*, 2005, **25**, 331-336.

EFFECT OF SWIRL ON SEPARATION PERFORMANCE OF A NEW SPRAY GRANULATION TOWER

M. Liu, Y. Mao, J. Wang and J. Wang

College of Chemical Engineering, China University of Petroleum, Beijing 102249, P.R.China.
Email: maoyu@cup.edu.cn

1 INTRODUCTION

A number of new techniques have been developed for efficient and clean utilization of heavy oil due to the increasing problems with natural resources shortage and environment deterioration. Among those, heavy oil stage separation is a technique using solvent deasphalting.[1] A major advantage of this process lies in that the solvent in asphalt phase is recovered at atmospheric pressure and low temperature through granulation of asphalt, which results in a simplified process flow and reduced investment.[2] An array nozzle distributor was introduced into a conventional cylinder-on-cone tower in order to meet the need of high solvent recovery rate in this process.[3] Experimental studies[4-6] demonstrated that particles of lower densities could be separated from the gaseous solvent more easily under the combined action of centrifugal and gravitational forces in the new granulation tower. In this work, the effect of swirl on separation performance is investigated using CFD technique, which has been widely used for engineering design and optimization.[7-10]

2 NUMERICAL METHODS

2.1 Physical Model

Figure 1 shows the configuration of the novel tower used in the experiment. The cylinder of the tower is 1.0 m in diameter and 2.50 m in height. The cone beneath the cylinder has a minimum diameter of 0.24 m and a height of 1.04 m. The dust-laden gas enters the tower through four nozzles of a distributor. These four nozzles can be rotated in vertical planes as shown in Figure 1b, allowing the direction of the phase entering the tower to be controlled. In addition, these four nozzles are located at the same level and the same distance (0.58 m) away from the centre of the tower, as shown in Figure 1c. The vertical distance between these nozzles and the top of the cylinder is 0.65 m. To collect particles separated from the gas, a dust hopper with a diameter of 0.70 m and a height of 0.85 m is connected to the cone.

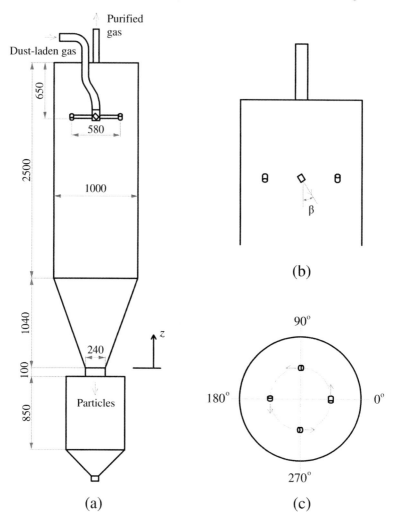

Figure 1 *Configuration of the tower. (a) Tower dimensions, (b) vertical inlet direction and (c) horizontal inlet direction.*

2.2 Numerical Simulation

The gas-solid flow is simulated using an Eulerian-Eulerian model based on the kinetic theory of granular flow, which allows the determination of the pressure and viscosity of the solid phase in place of empirical correlations.[11] A structured grid of 376,800 cells is used for the simulation. The effect of grid refinement was previously evaluated in the simulation process.[12] The finite volume method is used to solve the governing equations. These equations are discretized with the QUICK scheme. The phase-coupled SIMPLE algorithm is implemented to deal with the coupling between velocity and pressure. The solution is considered to be converged when the scaled residual of the continuity is below 1×10^{-4}.

All variables except pressure are specified at the inlet. An ambient pressure is given at the gas outlet, while appropriate velocity is designated at the particle outlet. The no-slip boundary condition is imposed at the wall. The near-wall flow is determined by the standard wall function.

3 RESULTS AND DISCUSSION

3.1 Model Validation

The calculated streamlines are compared with experimental observations, as shown in Figure 2. The gas jetted from nozzles spirals downward in the outer part of the tower. At the same time, it spirals upward with a relatively small diameter in the inner part and finally leaves through the outlet pipe located on the top of the cylinder. A qualitative agreement between the simulation and the experiment is obtained, as illustrated in Figure 2.

Figure 3 shows a quantitative comparison of separation efficiency and pressure drop between simulations and experiments. The swirl number, defined as the ratio of the tangential to the axial velocity, is set to be 1.00 for both simulations and experiments. The separation efficiency and pressure drop increase with the increase of gas flow rate, as shown in Figure 3. The numerical results agree well with the experimental data. The results shown in Figures 2 and 3 suggest that the numerical model used in this study can capture the gas-solid flow features in the novel tower.

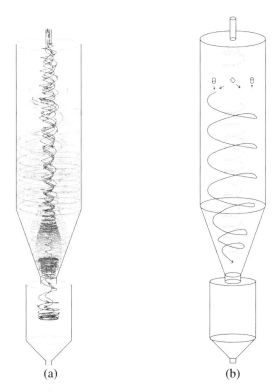

(a)   (b)

Figure 2   *Comparison of flow pattern. (a) Calculated streamlines, (b) experimental observations.*

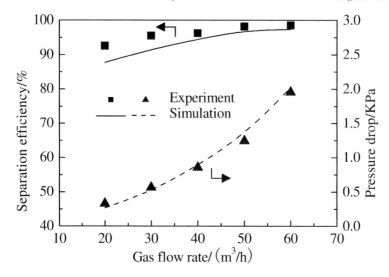

Figure 3 *Comparison of separation efficiency and pressure drop.*

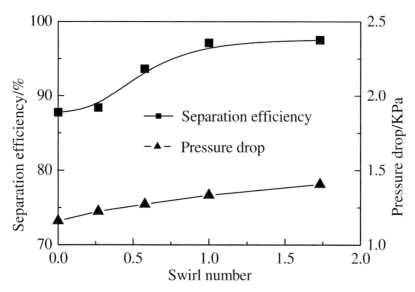

Figure 4 *Separation performance of the tower with different swirl numbers*

3.2 Effect of Swirl on Separation Performance

The calculated separation efficiency and pressure drop for different swirl numbers are given in Figure 4. A gas flow rate of 50 m³/h is specified for these simulations. It is clear that the separation efficiency increases with the increasing swirl number. For low swirl intensity, the separation efficiency increases slightly as the swirl number increases. As the swirl number increases further, the separation efficiency increases dramatically. When the swirl number exceeds 1.00, the enhancement of separation efficiency by swirl intensity becomes inconsiderable. Figure 4 illustrates that the pressure drop increases linearly with the increase of swirl number. There is an optimal range of swirl for the tower operation.

3.3 Effect of Swirl on Flow Field

In order to explore why separation efficiency increases with the increasing swirl, the flow fields in the novel tower with different swirl numbers are examined. Figure 5 shows the tangential velocity distribution in the tower with different swirl numbers. Although the tangential velocity profiles have a similar pattern, the magnitude of them is quite different. Obviously, tangential velocity in the new granulation tower increases significantly when the swirl number is increased from 0.00 to 1.73. A similarity in variation of tangential velocity is found for different axial and azimuthal locations. Higher tangential velocity leads to larger centrifugal force, which in turn results in higher separation efficiency.[13] That is why the separation efficiency increases with increasing swirl number.

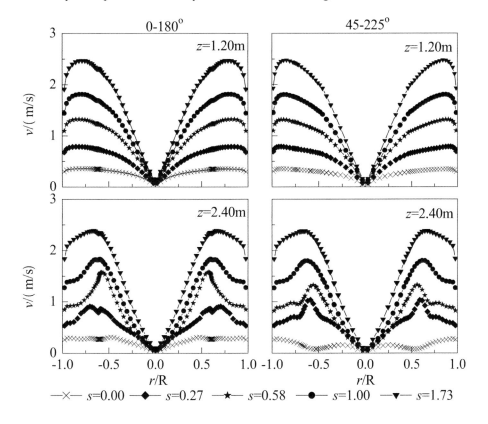

Figure 5 *Tangential velocity distributions for different swirl numbers*

Figure 6 illustrates the effect of swirl number on the distribution of solid concentration. A substantial decrease in the solid volume fraction is observed in the inner part of the tower when the swirl number is increased from 0.00 to 1.73. Simultaneously, the solid volume fraction increases in the annular region very close to the wall. This indicates that the increasing swirl intensity accelerates the outward migration of particles. A similar trend is observed for all axial positions. Furthermore, due to this outward migration, the solid volume fraction decreases along the negative axial direction in the tower except for the region very close to the wall.

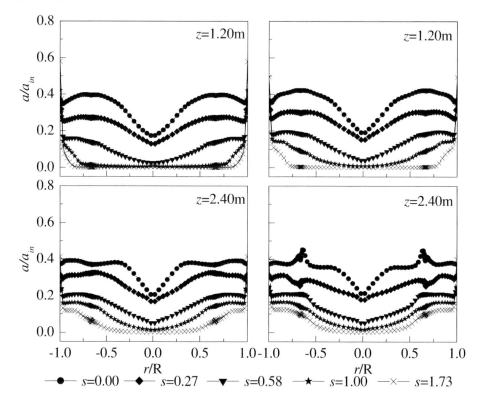

Figure 6 *Profiles of solid volume fraction for different swirl numbers*

4 CONCLUSIONS

A numerical study is carried out to investigate the effect of swirl intensity on separation performance of a new spray granulation tower. The predicted flow pattern is in qualitative agreement with the experimental observation, which consists of an outer downward spiralling flow and an inner upward one. In addition, the calculated separation efficiency and pressure drop agree well with the experimental data, providing a further validation of the numerical model. The simulation results for different swirl numbers show that the tangential velocity increases with increasing swirl number, resulting in an amelioration of separation efficiency, while the pressure drop is less affected by the swirl number.

References

1 C.M. Xu, S.Q. Zhao, X.W. Sun and Z.M. Xu, *CIESC Journal*, 2010, **61**, 2393.
2 S.Q. Zhao, C.M. Xu, R.A. Wang, Z.M. Xu, X.W. Sun and K.H. Chung, *US 7597794B2*, 2009.
3 J. Wang, Y. Mao and J.Y. Wang, *CN 2008101676764*, 2008.
4 J. Wang, Y. Mao, M.L. Liu and J.Y. Wang, *6th Annual Conference of Chinese Society of Particuology*, 2008, 895.
5 B. Li, *Master thesis*, China University of Petroleum-Beijing, 2008.
6 J. Wang, Y. Mao, X.W. Sun and J.Y. Wang, *CIESC Journal*, 2011, **62**, 393.
7 D.J.E. Harvie, T.A.G. Langrish and D.F.Fletcher, *Chem. Eng. Res. Des.*, 2001, **79**, 235.
8 T.A.G. Langrish, J. Williams and D.F. Fletcher, *Chem. Eng. Res. Des.*, 2004, **82**, 821.
9 J.W. Lee, H.J. Yang and D.Y. Lee, *Powder Technol.*, 2006, **165**, 30.
10 J. Cui, X.L. Chen, X.Gong and G.S. Yu, *Ind. Eng. Chem. Res.*, 2010, **49**, 5450.
11 H.L. Lu and D. Gidaspow, *Ind. Eng. Chem. Res.*, 2003, **42**, 2390.
12 M.L. Liu, Y. Mao, J.Y. Wang and J. Wang, *6th Joint China-Japan Chemical Engineering Sympostium*, 2011, 28.
13 R.B. Xiang and K.W.Lee, *Chem. Eng. Process.*, 2005, **44**, 877.

COMBINED EFFECTS OF ACOUSTIC WAVE AND GAS-SOLID JET ON AGGLOMERATION OF INHALABLE PARTICLES FROM COAL COMBUSTION PLANT

Z. YANG and Q. GUO

Key Laboratory of Clean Chemical Processing Engineering in Universities of Shandong, College of Chemical Engineering, Qingdao University of Science Technology, Qingdao, Shandong 266042, China. Email: qj_guo@yahoo.com.

1 INTRODUCTION

The emission of coal-fired inhalable particles, especially those with an aerodynamic diameter less than 2.5 μm (PM2.5), as a serious environmental issue, have attracted much attention. The evidence on inhalable particles and public health demonstrates adverse health effects at exposures experienced by urban populations in cities throughout the world.[1] Therefore, the removal of inhalable particles is one of the key problems in the field of environmental protection. Numerous particle emissions are produced from coal combustion. In general, particles greater than 10 μm can be captured by conventional devices such as electrostatic precipitators, cyclones and wet scrubbers. However, the capture efficiency of filters for PM2.5 drops significantly, quantities of PM2.5 particles are emitted into the ambient air.[2, 3, 4]

In this paper, a turbulence field generated by gas-solid jet flow ejected into an agglomeration chamber combines with acoustic field to intensify particle collision. Combined intensification effects on particles removal are explored.

2 EXPERIMENTAL

The coal-fired fly ash particles collected in the electrostatic precipitator from Qingdao power plant was used in this study. The density of fly ash was 1838 kg/m³ with the moisture content 0.78%. As shown in Figure 1, the experimental setup was composed of a fluidized bed aerosol generator, a gas-solid jet system, an acoustic generator, a measurement apparatus，and a particles agglomeration chamber. The agglomeration chamber was made of a transparent polymethyl methacrylate cylindrical tube with an inner diameter of 116 mm and 600 mm in length. A signal generator (WuYi WY32003) generated sound wave to transmit an amplifier located at one side of the agglomeration chamber. Air was filtered by a High-Efficiency Particulate Air (HEPA) filter to pass through the fluidized bed aerosol generator. The fly ash particles in small size were subsequently entrained in gas flow to enter the agglomeration chamber. Depending on the experiment requirements, the other gas flow was introduced into the humidifier to achieve experimental relative humidifier, and passed through fluidized bed dispersing generator to yield gas-solid flow mixing with coarse particles, ejected into the agglomeration chamber by injection nozzle.

To achieve a better aggregation effect, experimental sound pressure level were kept at 120 dB.[5] By using an iso-kinetic sampling probe, fine particles were distributed into the cascade impactor. Table 1 described the cut point diameters of a cascade impactor.

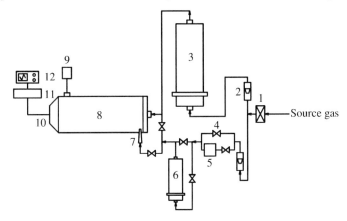

Figure 1 *Schematic diagram of the inhalable particle agglomeration apparatus*

1.	HEPA filter	2.	Rotameter
3.	Fluidized bed aerosol generator	4.	Valve
5.	Humidifier	6.	Fluidized bed disperse generator
7.	Injection nozzle	8.	Agglomeration chamber
9.	Sampling instrument	10.	Horn
11.	Power amplifier	12.	Signal generator

Table 1 *Cut point diameters of the cascade impactor*

Stage	Particle diameter /μm	Pore diameter/mm
0	Submicron	Membrane
1	0.43~0.65	0.25
2	0.65~1.1	0.25
3	1.1~2.1	0.34
4	2.1~3.3	0.53
5	3.3~4.7	0.71
6	4.7~5.8	0.91
7	5.8~9.0	1.89
8	9.0~10	2.25

3 RESULTS AND DISCUSSION

3.1 Particle Agglomeration Characteristics in the Combined External Field

In the agglomeration chamber, coal-fired fly ash inhalable particles were excited by acoustic wave and gas-solid jet flow. The particle mass distribution curves with and without external field are depicted in Figure 2. A dramatic reduction in inhalable particles mass is observed, indicating that the small particles form larger agglomerates. The sample masses of initial distribution, acted by acoustic wave, gas-solid jet and combined process are 110.63, 85.34, 80 and 62.85 mg, respectively. This implies the removal efficiencies of fine particles are 22.86% and 27.69% by acoustic wave and gas-solid jet action, moreover, the combined application approaches a maximum removal efficiency of 43.19%. It can be concluded that combined application will greatly enhance the performance of the filtering device installed downstream.

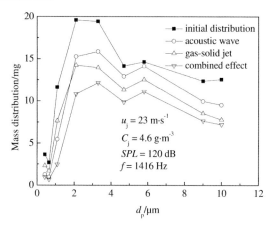

Figure 2   *Mass distribution of particle agglomerates exposed in external field*

3.2 Effect of Coarse Particles in Jet on Particle Agglomeration

Figure 3 shows the influence of coarse particle concentration in jet on the removal efficiency. Note that the mass removal efficiency of combined process without coarse particles is only 36.24%. By adding coarse particles ranging from 96 μm to 120 μm, the removal efficiency increases to 42%. In contrast, as coarse particle size is increased from 120 to 250 μm, an efficiency of 47% is obtained. There are different agglomeration effects of inhalable particles with adding coarse particles of various sizes, and large particles are able to strengthen turbulence intensity with large flow region.[6]

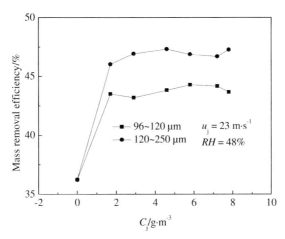

Figure 3 *Effect of agglomeration kernel concentration in jet on particle removal efficiency*

3.3 Effect of Sound Frequency on Particle Agglomeration

Figure 4 describes the effect of sound frequency on particle mass removal efficiencies. The increasing acoustic frequency results in an initial increase in removal efficiency until the efficiency reaches the peak value, and then a decreasing efficiency is observed. The optimal removal efficiency is found at 1,416 Hz.

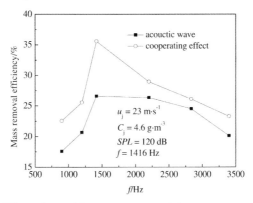

Figure 4 *Effect of sound frequency on particle mass removal efficiency*

In gas-solid jet flow, gas turbulence provides energy for inhalable particles to strengthen the collision, consequently, causes agglomeration. At the acoustic field at the frequency of 1,416 Hz, the relative entrainment of acoustic wave on particle pairs reaches a maximum. In the combined process, acoustic wave and gas-solid jet promote at the optimal frequency, the particle removal efficiency is hence improved significantly. For industrial applications characterized by poly-disperse aerosols, this combined process is beneficial due to their low attenuation loss through aerosols.

3.4 Effect of Jet Velocity on Particle Agglomeration

The Effect of jet velocity on particle agglomeration is illustrated in Figure 5. An increase in jet velocity initially results in an increase in the removal efficiency at a given experimental conditions, and then causes a decreasing efficiency. For the combined effects, the reduction rate of removal efficiency is lower than that of the separate jet action. The maximum removal efficiency occurs when the jet velocity is 25 m·s^{-1}.

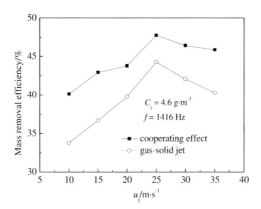

Figure 5   *Effect of St in jet exit on particle agglomeration*

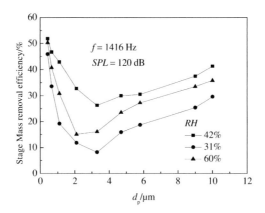

Figure 6   *Staged mass removal efficiency of inhalable particles at different RH*

3.5 Effect of Relative Humidity on Particle Agglomeration

Figure 6 shows the staged removal efficiency of inhalable particles in different *RH*. At 42% *RH*, all levels staged removal efficiency reach the maximum value. There is a great mass removal efficiency for small particles ($d_p < 2$ μm). For large particles ($d_p > 9$ μm), as the contact area with water vapor grows, moisture content among particles increases, consequently, capillary effect becomes significant. Furthermore, a high probability is achieved for large particles to collide other particles, making easy to form large aggregates. Removal efficiency for coarse particles is lower than fine particles.

4 CONCLUSIONS

(1) The adoption of coarse particles in jet shows its effectiveness for inhalable particle agglomeration.

(2) With or without acoustic wave, there is an optimum frequency for given inhalable particles to reach the best agglomeration efficiency.

(3) The removal efficiency initially increases, and then starts to reduce once it reach a maximum value. For combined effects, the downtrend of removal efficiency is lower than that of separate jet action.

(4) Considering different adhesion mechanisms among inhalable particles under different RH, there is a high mass removal efficiency for small particles ($d_p < 2$ μm) and large particles ($d_p > 9$ μm).

Nomenclature

C_j	—	jet fly ash concentration, g·m^{-3}	RH	— relative humidity, %
d_p	—	particles diameter, μm	SPL	— sound pressure level, dB
f	—	frequency of sound wave, Hz	u_j	— jet gas velocity, m·s^{-1}

References

1   World health organization, *WHO air quality guidelines global update 2005*, Bonn, Germany, 2005.
2   J.A. Gallego-Juárez, E. R.-F. De Sarabia, G. Rodríguez-Corral, T. L. Hoffmann, J. C. Gálvez-Moraleda, J. J. Rodríguez-Maroto, F. J. Gómez-Moreno, A. Bahillo-Ruiz, M. Martín-Espigares and M. Acha, *Environ. Sci. Technol.*, 1999, **33**, 3843.
3   L. Yang, J. Yan and X. Shen, *Mod. Chem. Ind.*, 2005, **25**, 22.
4   S. Heidenreich, U. Vogt, H. Büttner and F. Ebert, *Chem. Eng. Sci.*, 2000, **55**, 2895.
5   D. Sun, PhD Thesis, Qingdao University of Science Technology, 2009.
6   X. Yan and X. Wang, *Journal of Engineering Thermophysics*, 2008, **29**, 96.

THE EFFECT OF CYCLONE HEIGHT ON FLOW FIELD AND SEPARATION PERFORMANCE

C. Gao, G. Sun, J. Yang, R. Dong and J. Liu

State Key Laboratory of Heavy Oil Processing, China University of Petroleum, Beijing, 102249, China

1 INTRODUCTION

Cyclone separator (CS) is a key component for reducing dust contents in many advanced processes, such as coal combustion, gasification, petroleum refining and oil shale retorting. Its performance is mainly determined by the structural dimension. In the past, the effects of some cyclone dimensions, such as gas inlet and outlet dimension, were extensively investigated. However, the effect of cyclone length receives less attention and is less well understood. Most models have predicted that an increase in length improved the separation efficiency.[1,2] However, these were not fully validated. According to Barth[3] and Mothes and Loffler,[4] a less intensive vortex means less centrifugal force on the particles in CS. Conversely, the radial velocity across a longer CS will be less, so that the inward drag on the particles is also reduced. It cannot be predicted in advance what the effect will be. When considering the time-of-flight models, similar problems also exist. Recently, the effect of cyclone height on the performance has been investigated by many researchers.[5-9] Shi and Wang[10] found that, at the same velocity, larger height-diameter ratio H/D results in a longer mean residence time of particle and higher efficiency. However, the height-diameter ratio has an optimal value of 3~3.5. Further increase of the ratio (say >3.5) the separation efficiency is reduced. Similar results were obtained by Jin,[11] who showed that the optimal height is 2.8D~3.2D. Moreover, Hoffmann[12-14] found that a marked improvement in cyclone performance with increasing length is noticed. But, the efficiency reduces dramatically when the length is increased to certain value. Hoffmann attributed the dramatic fall in separation performance to the 'natural turning' phenomenon. Nevertheless, no detail analysis was given. Our current knowledge cannot provide clear instructions for practical cyclone design. Revealing the effect of cyclone height on the flow field is still a crucial problem to be solved. In this paper, Quantitative analysis of the effect of cyclone height on flow structure and separation performance is performed aiming to provide insights into the design optimization of cyclones.

2 NUMERICAL SIMULATION METHOD

The geometrical dimensions of the cyclone separator considered are depicted in Figure 1.

The diameter of the cyclone is 186 mm, and the inlet size is 110 mm x 45 mm. Due to the diversity of cyclone structure, the contributions to separation performance from cylinder and cone section are different and need to be examined separately. When investigating the effect of cone height, the cylinder height is fixed as $h = 1D$ (i.e. 170 mm). Similarly, when investigating the effect of cylinder height, the cone height is fixed as $H_c = 3D$ (i.e. 560 mm). Various cyclone structures considered are shown in Table 1.

Table 1 *Cyclone height (unit: mm)*

	Varying cone structure					Varying cylinder structure			
	1D1D	*1D3D*	*1D4D*	*1D4.5D*	*1D5D*	*1D3D*	*2D3D*	*3D3D*	*5D3D*
Cylinder	170	170	170	170	170	170	350	520	910
Cone	186	560	740	840	930	560	560	560	560

Structured mesh is used, as shown in Figure 2. The origin of the coordinates is set at the bottom end of the vortex finder, and the positive direction is downward. RSM in the platform of Fluent 6.1 is used. The pressure-velocity coupling algorithm SIMPLEC and the QUICK higher order upwind interpolation scheme are used in all numerical simulations. The gas inlet velocity is set as $V_i = 20$ m/s. The particle used is 800 mesh talcum powder with a density of 2700 kg/m^3, and the inlet solids concentration C_i is 10 g/m^3. The solid concentration distribution can be calculated using the Discrete Phase Model (DPM). The particle velocity is equal to the gas inlet velocity.

The boundary conditions at the outlet of the cyclone are prescribed as a fully-developed pipe flow. At the top end of the gas outlet, the gradients of all variables in the axial direction are assumed to be zero. No-slip condition is assumed at the wall. For the grid nodes near the wall, they are approximated and treated using the wall function.

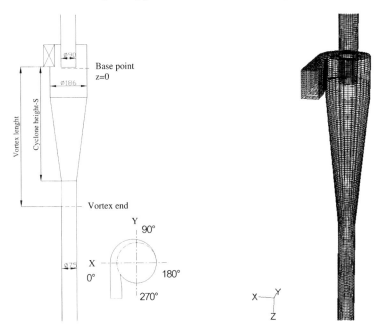

Figure 1 *Separator structure*   **Figure 2** *The sketch of computation grids*

3   RESULTS AND DISCUSSION

3.1   Effect of Cyclone Height on the Flow Field

3.1.1   The tangential velocity. The cyclone height has an important effect on the swirl intensity. When increasing the cyclone height, the shell wall area increases, leading to additional friction and lower swirl intensity. The tangential velocity profiles at section z=60 mm and z=160 mm with various cone heights are shown in Figure 3. It is clear that the tangential velocity profile can be described as a Rankine vortex, a combined free and forced vortex. As cone height increases, the position of maximum tangential velocity, namely the interface between inner and outer vortex, is almost invariant. However, the tangential velocity gradient of outer vortex and the maximum tangential velocity decrease continuously. At section z=60, the maximum tangential velocity of cyclone 1D1D is $2.35v_{in}$. When the cyclone is enlarged to 1D5D, the maximum tangential velocity is $1.81v_{in}$, which is only 77.29% of 1D1D's. For various cylinder heights, the tangential velocity has a similar pattern as shown in Figure 4. However, the gas swirl intensity decreases as the cylinder height increases. In a cyclone, the inertia force arising from gas swirl affects mainly particle separation. The decrease in the tangential velocity may lead to poorer separation performance.

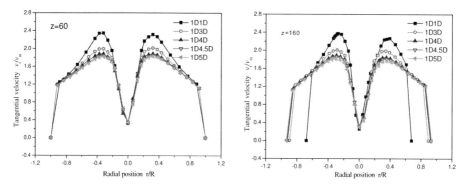

Figure 3   *The tangential velocity profiles for various cone heights*

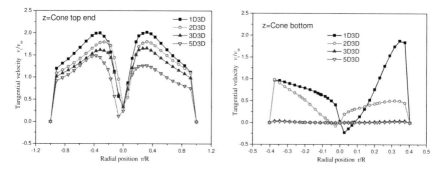

Figure 4   *The tangential velocity for various cylinder heights*

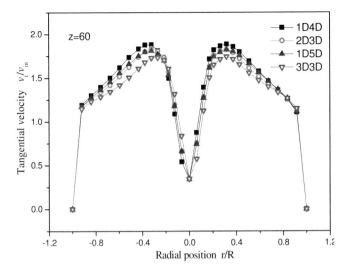

Figure 5 *Tangential velocity profiles at z=60 for various cyclones*

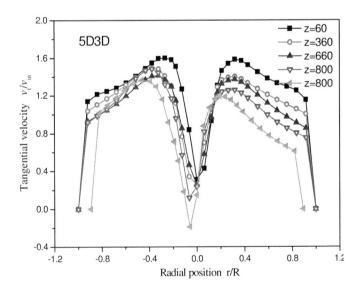

Figure 6 *Tangential velocity profiles of 5D3D at various positions*

Further analysis indicates that, under the same total height (1D4D / 2D3D，1D5D / 3D3D), a longer cone structure results in a higher tangential velocity (Figure 5). This is attributed to that the tangential velocity is already attenuated in the cylinder section when the cylinder height is increased (Figure 6). On the other hand, as the cone height increases, the maximum tangential velocity attenuates less, due to the restriction of the cone structure

(Figure 7). Obviously, a longer cone structure is more beneficial to compensate friction.

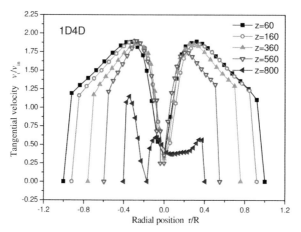

Figure 7   *Tangential velocity profile of 1D4D at various positions*

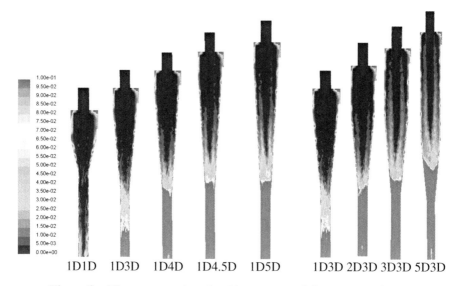

Figure 8   *The concentration cloud images at x=0 for various cylcones*

3.1.2   The vortex length. Except its effect on swirl intensity, an increase in cyclone height will increase the effective travel distance of the swirling gas. This is an important factor that cannot be neglected when investigating the effect of cyclone height. The concentration cloud images are presented in Figure 8. In a short cyclone (1D1D), the particles flow down to the cyclone bottom in streamers following a helical path. With the increase of cyclone height, the particles begin to concentrate in dip leg, and result in wall deposits. It suggests that the swirling strength has attenuated rapidly. This phenomenon is similar to that reported by Hoffmann.[14]   Thus, it is considered that the ring indicates the position of vortex end. In addition, Hoffmann[14] argued that, when the physical length is longer than the vortex length, it entails poor collection performance, and the space under the vortex end is wasted. The effective separation space is the gap between the bottom of vortex finder and the vortex end, namely vortex length.[15] The vortex lengths for different cone and cylinder heights obtained from numerical simulations are shown in Figures 9 and 10. The vortex length increases as the cone or cylinder height increases, but at a lower rate. The vortex end moves up to the cone bottom. When the total height is 5D (1D4D,2D3D), the vortex end enters into the cone.

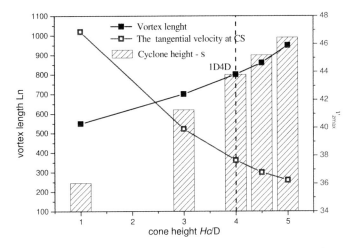

Figure 9 *Positions of vortex end with different cone heights*

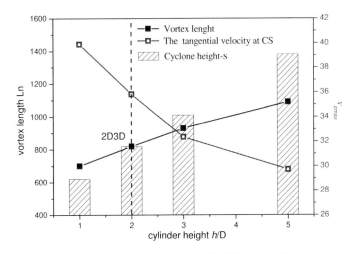

Figure 10 *Positions of vortex end with different cylinder heights*

Figure 11 shows the concentration profiles with various cone heights. With a short cone (1D1D), there is not enough time for the particles in upward flow to be separated. The small particles escape directly through vortex finder, resulting in a high solid concentration in the inner vortex. As the cone height increases, the solid concentration in the inner vortex decreases. However, when the cyclone height increases to 1D4D, the vortex end enters into the cone and the solid concentration at CS section increases. For instance, at the cone bottom, the solid concentration at the center of cyclone 1D5D is 0.52 kg/m^3, and reaches to 52 times of the inlet concentration. Thus, it is believed that the appearance of vortex end in conical section leads to severe particle back mixing and affects the cyclone performance significantly. When the cylinder height is increased, the concentration profiles exhibit similar characteristic, as shown in Figure 12.

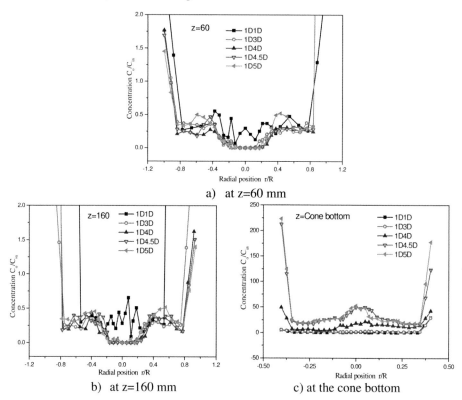

Figure 11   *The concentration profiles with various cone height*

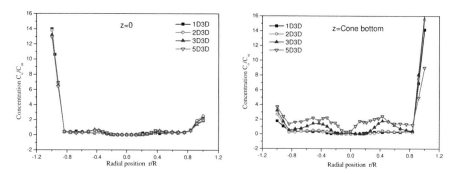

Figure 12   *The concentration profile as a function of cylinder height*

It is clear that the cone height and the cylinder height have similar effects on cyclone flow field (Figures 9 and 10). As the cyclone height increases, the tangential velocity decreases and the particle separation efficiency decreases. At the same time, the separation space is increased, inducing longer particle residence time and higher separation probability. In addition, the existence of vortex end increases the complexity of flow field.

3.2 Effect of Cyclone Height on the Separation Efficiency

Figures 13 and 14 show the effect of cyclone height on the separation efficiency. It is obvious that the separation efficiency is improved with the increasing cone height, and the maximum efficiency is obtained with 1D4D, while the efficiency reduces when the total height exceeds 5D. When the cylinder height is increased, an optimal cyclone height is still obtained. The cyclone 2D3D has the best separation efficiency of 92.44% (The optimal total cyclone height is 5D). It can be inferred that the ratio of cone and cylinder height has less effect on the optimal height.

In view of the flow characteristic of cyclone, it is believed that there are two mechanisms that the cyclone height affects the performance of cyclone. When the cyclone is short, the re-entrained small particles cannot be separated secondarily and escape directly through vortex finder, leading to the reduction in separation efficiency. As the cyclone height increases, the maximum tangential velocity decreases, and cyclone performance is reduced, while the increased separation space leads to longer particle residence time and higher separation probability. This is the reason for improved performance in a certain cyclone height range. With a certain cyclone height, the vortex end enters into the cone, and severe particle back-mixing occurs, resulting in the reduction in separation efficiency. Thus, the design of optimal cyclone height must take the vortex length into account.

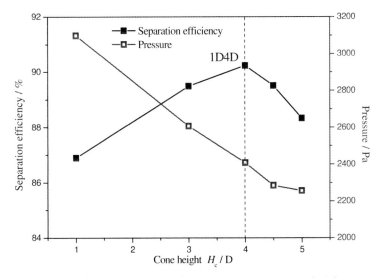

Figure 13 *Separation performance at various cone heights*

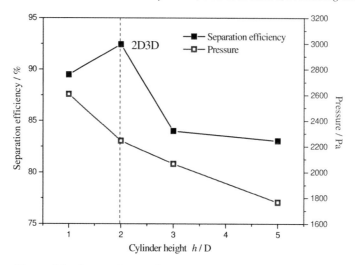

Figure 14 *Separation performance at various cylinder heights*

4 CONCLUSIONS

In order to quantify the effect of cyclone length on performance, the flow field and efficiency of cyclone with various heights are analysed numerically. It has been found that

(1) The cone and cylinder heights have similar impact on the cyclone flow field and performance, but the ratio of cone to cylinder height only has limited effect. Moreover, With the same total height, the maximum tangential velocity of long cylinder structure has already attenuated in cylinder section and its swirl intensity is slightly reduced as the cone height increases.

(2) The separation efficiency increases with the increasing cyclone height and reaches maximum when the cyclone height is up to a certain value. An optimal cyclone height is hence obtained. It is anticipated that, in a short cyclone, although the gas flow has high swirl intensity, the re-entrained small particles cannot be separated secondarily due to the short separate space, and escape directly through vortex finder, leading to the reduction in separation efficiency. When the cyclone height increases, the tangential velocity decreases. However, the increased separation space leads to longer particle residence time and higher separation probability. When the cyclone height is up to certain value, the vortex end enters into the cone, and results in the drop of efficiency.

Acknowledgements

The authors acknowledge the financial support from National Basic Research Program of China (Grant 2005CB22120103)

References

1   D. L. Iozia and D. Leith. *Aerosol Sci. and Technol.* 1989, **10**:491.
2   D.Leith and W. Licht. *AIChE Symp.Ser.*. 1972, **68**, 196.
3   W. Barth. Berechnung und Auslegung von Zyklonabscheidern auf Grund neuerer Unterscuhungen. Brennst-Waerme-Kraft, 1956,**8**:1-9.

4   H. Mothes and F. Loffler. *Int. Chem. Eng*. 1988, **28**, 63.
5   J. Zhang and Y. Jin. *PETRO-CHEMICAL EQUIPMENT*, 2007, **36** , 33.
6   L. Wang, Z. Hao and X.Wang. *Journal of engineering thermophsics* , 2009, **30,** 223.
7   R.B. Xiang and   K.W. Lee. *Chemical Engineering and Processing*. 2005, **44**, 877.
8   H. Safikhani, M.A.Akhavan-Behabadi and M. Shams. *Adv. Pow. Techn*. 2010,**21**,435.
9   Y. Zhu and K. W. Lee. *Aerosol Sci*. 1999, **30,** 1303.
10   M. Shi and Y. Wang. *Petro-Chemical Equipment Technology*, 1992, **13**, 14.
11   G. Jin. *Chemical equipment design*. Beijing: Chemical industry press, 2002.
12   A. C.Hoffmann, M. de Groot and W. Peng. *AIChE Journal*. 2001, **47**, 2452.
13   A. C. Hoffmann, A. V. Santen and R. Allen. *Powder Technol.,* 1992. **70**, 83.
14   A. C. Hoffmann, R. de Jonge and H. Arends. *Filt. Sep.*, 1995, **32,** 799.
15   R. M. Alexander,   *Proc. Austral. Inst. Min. Met.*, 1949, **152**, 203.

INFLUENCE OF FLUIDIZATION CONDITIONS ON STICKING TIME DURING REDUCTION OF Fe_2O_3 PARTICLES WITH CO

B. Zhang, [1,2] Z. Wang,[2] X. Gong[2] and Z. Guo[1,2]

[1] State Key Laboratory of Advanced Metallurgy, University of Science and Technology Beijing, Beijing 100083, China. E-mail: zcguo@metall.ustb.edu.cn
[2] National Engineering Laboratory for Hydrometallurgical Cleaner Production Technology, Institute of Process Engineering, Chinese Academy of Sciences, Beijing 100190, China

1 INTRODUCTION

Due to no need of prior treatment, excellent heat and mass transport, fluidized beds were selected as reactors for reduction or pre-reduction of iron ores as an alternative to blast furnace.[1-4] However, the industrialized development of the fluidized bed in ironmaking was limited,[5-7] which was mainly attributed to the defluidization resulted from sticking among iron ore particles.[8-11]

The sticking phenomenon has been extensively investigated and summarized by many researchers. The sticking temperature increased with the reduction in gas velocity.[9,12,13] The sticking tendency depended strongly on the temperature. The sticking was not observed at a temperature below 873 K.[9,13] And the sticking tendency increased for small particles with a low kinetic energy.[9] Hayashi *et al.*[10] found that spheroid shape of iron ores had minor tendency to sticking, compared to angular shape. Komatina and Gudenau[9] proved that the sticking tendency of ores with higher amount of gangue was smaller. Stephens and Langston[14] claimed that the sticking was directly proportional to adhesive force and contact area among fine ore particles, and inversely proportional to their momentum. Nevertheless, an appropriate parameter to describe the sticking was not provided in literature, so the quantitative relationship between the sticking and the factors was not clear.

The reduction degree of iron ores and the iron whisker were considered as the key factors on the sticking in earlier work.[9-12] However our previous study proved that the sticking of pure iron particles without the iron whisker occurred at a temperature of approximately 973 K.[15] It was found that the sticking was strongly influenced by the metallic iron precipitated on the particle surface.[16] Thus, based on the iron precipitation, the mechanism of the sticking was proposed. However, the influence of fluidization conditions on the sticking from characteristics of precipitated iron was scarcely investigated.

To simplify the chemical composition of iron ores, chemical grade Fe_2O_3 particles were selected as the model material for reduction in the fluidized bed. The sticking time based on the pressure drop was defined to characterise the sticking. And the effect of fluidization conditions on the sticking time was investigated. The underlying mechanism was discussed.

2   EXPERIMENTAL

2.1   Experimental Setup

Using the experimental apparatus and preparation process of Fe_2O_3 particles reported by Zhang *et al.*,[16] batches of Fe_2O_3 particles were fed into the fluidized bed reactor, and then fluidized with N_2 at a flow rate of 3 L/min. The fluidized bed was heated to the specified temperature. After maintaining the system at the specified temperature for a few minutes to keep N_2 wash the system so it is free of air, N_2 was switched to the reduction mixture gas. The experimental conditions were shown in Table 1. The superficial gas velocities U were 5 times greater than those at minimum fluidization U_{mf}, indicating that the gases had the same kinetic energy for stirring a bed. Once the complete defluidization occurred and the pressure was steady, the reduction mixture gas was switched back to the pure N_2 at a flow rate of 1 L/min immediately, as shown in Figure 1. The atmosphere was kept until the temperature of the particles dropped to room temperature. Finally, the particles were obtained from the reactor and kept in the dryer for further characterisation.

Table 1 *Experimental conditions*

Number	Temperature / K	Reduction mixture gas / L/min			U/U_{mf}
		CO	CO_2	N_2	
1		2.97	0	0	5
2		2.67	0	0.27	5
3		2.38	0	0.54	5
4		2.08	0	0.81	5
5	1073	2.67	0.24	0	5
6		2.38	0.49	0	5
7		2.08	0.73	0	5
8		1.78	0	0	3
9		2.37	0	0	4
10		3.56	0	0	6
11	1173	2.42	0	0	5
12	1123	2.70	0	0	5
13	1023	3.13	0	0	5

After each test, a blank test was carried out to examine the blank pressure. The pressure drop of the powder layer was equal to the actual pressure minus the pressure recorded in the blank test. The measurement of the metallization ratio of particles was introduced in Zhang *et al.*[16] The specific surface area of particles was determined from nitrogen gas adsorption-desorption (Quantachrome, Q UADRASORB SI-MP) at the boiling point of liquid nitrogen (i.e. 77.4 K).

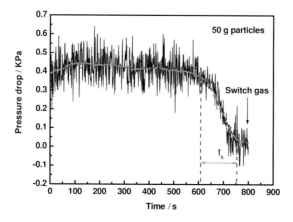

Figure 1   *Variation of the pressure drop with reduction time*

2.2 Definition of the Sticking Time

The sticking during the reduction in the fluidized bed was determined by combining the abrupt change of pressure drop though bed with the observation of the state of fluidization. Figure 1 shows the variation of the pressure drop with the reduction time reduced by CO-N_2 gas mixture (CO 2L/min, N_2 1L/min) at 973 K. At the initial stage of reduction, the pressure drop was steady and the state of particles was in fluidization. Later on, the pressure drop fell sharply and the state of particles was in defluidization. The sticking time (t_s) is defined from the occurrence of sticking (the start of the dramatic drop of the pressure) to the complete defluidization (the finish of the dramatic drop of the pressure). As shown in Figure 1, the time between the two intersections of three tangent lines is the sticking time. A longer sticking time indicated smaller sticking tendency.

3 RESULTS AND DISCUSSION

3.1 Effect of CO Concentration

In order to examine the influence of CO concentration on the sticking time, 30 g Fe_2O_3 particles were reduced with CO concentration from 70% to 100% in the fluidized bed, shown as Test No. 1-4 in Table 1. Figure 2 presents variations of the sticking time and the metallization ratio with CO concentration. The sticking time is 59 s with 70% CO gas. It decreases with increasing CO concentration, and the time is 46 s with 100% CO gas. The metallization ratios of particles under 70%, 80%, 90% and 100% CO gas reach 17.91%, 19.26%, 19.60% and 22.15%, respectively. The metallization ratio increases with CO concentration due to the chemical kinetics of gas-solid reaction. It indicated that the sticking time decreased with increasing CO concentration, so did the metallization ratio.

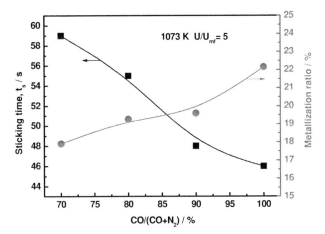

Figure 2 *Variations of sticking time and metallization ratio with CO concentration*

In our previous work,[16] the precipitated iron was both on the surface and inside of the reduced particles where the sticking occurred. The similar phenomenon was observed during the reduction of iron ores with H_2.[17] The iron layer covering the particle was not formed, but the metallic iron with the grain precipitated on the surface. This indicated that the metallization ratio signified the probability of iron-iron contact, thereby it is necessary to estimate the contact area of precipitated iron when particles collide together. It was found that the sticking time decreased with increasing iron-iron contact area determined by the metallization ratio.

3.2 Effect of Reduction Potential

30 g Fe_2O_3 particles were reduced with reduction potential ranging from 70% to 100%, (Test No. 1, 5-7 in Table 1), to explore the influence of the reduction potential on the sticking time. Figure 3 shows variations of the sticking time and the metallization ratio with reduction potential. The sticking time with reduction potential of 70% is 85 s and higher than that with reduction potential of 90% or 100%. This is due to a low metallization ratio of 10.90%, compared to the results in Figure 2. When the reduction potential is over 80%, the metallization ratios are almost identical, whereas the sticking time was gradually reduced from 94 s to 42 s, which is different from the results shown in Figure 2.

Previous studies[9, 18-20] of product morphology showed that reduction potential affected the structure of particles. For this reason, the specific surface area of reduced particles was measured to explore the relationship between the sticking time and the structure of particles, as shown in Figure 4. The specific surface area of raw material is 0.065 m^2/g. The specific surface area of reduced particles with reduction potential of 80%, 90% and 100% are 0.233 m^2/g, 0.235 m^2/g and 0.278 m^2/g, respectively, revealing that the specific surface area of particles before and after reduction changes greatly and that increases with increasing reduction potential. This leads to higher probability of iron-iron contacts when particles collide together. Therefore, the iron-iron contact area increased with increasing reduction potential, due to the structure change of particles. The increase in specific surface area of particles affected by reduction potential results in a decrease in sticking time, as a consequence of the increase in iron-iron contact area.

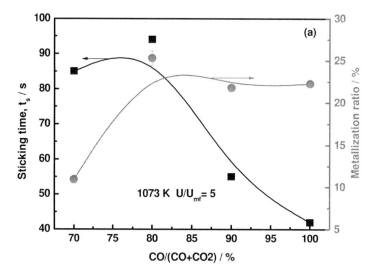

Figure 3　　*Variations of sticking time and metallization ratio with reduction potential*

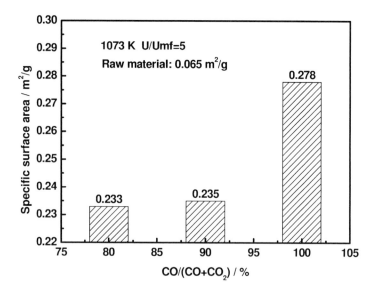

Figure 4　　*Variations of specific surface area with reduction potential*

3.3 Effect of Gas Velocity

To investigate the influence of gas velocity on the sticking time, 30 g Fe_2O_3 particles were reduced with gas velocity from 3 U_{mf} to 6 U_{mf}, (see Test No. 1, 8-10 in Table 1). Variations of the sticking time and the metallization ratio with gas velocity are shown in Figure 5. The sticking times at 3 U_{mf} and 6 U_{mf} are 38 s and 51 s, respectively, and it increases with gas velocity. With the increase of gas velocity, the metallization ratio gradually raises from 12.28% to 23.10%. This trend is opposite to that shown in Figure 2, implying that the sticking time was influenced by gas velocity, and which could outbalance that of the metallization ratio.

The gas flow passed through the powder layer in the fluidized bed and generated drag force on the particles. The drag force increased with gas velocity.[21] For the same particles, larger drag force induced higher momentum of particles, which led to less sticking. Gudenau *et al.*[22] reported that the sticking temperature increased with gas velocity. On the other hand, the sticking tendency decreased with increasing particle size in that particles had higher momentum. Degel[13] found that the sticking appeared for ore Carajas of particle size 60-90μm at temperature 1073 K, and of the size 90-150μm at temperature 1173 K. This illustrated that the sticking time increased with gas velocity due to the increasing momentum of particles.

3.4 Effect of Temperature

In order to examine the influence of temperature on the sticking time, 30 g Fe_2O_3 particles were reduced at temperature from 1023 K to 1173 K. Figure 6 shows variations of the sticking time and the metallization ratio with temperature. The sticking times at 1023 K, 1073 K, 1123 K and 1173 K are 47 s, 45 s, 41 s and 31 s, respectively. The metallization ratio decreased from 23.51% to 10.73% with increasing temperature. Contrary to the results shown in Figure 2, the sticking time and the metallization ratio decreased with increasing temperature, implying that it was influenced by temperature, which could outweigh that of the metallization ratio.

Our previous study[15] reported that the sticking of pure iron particles fluidized by N_2 did not occur until the temperature rose to 973 K. For pure iron particles, the metallization ratio was 100%, implying the constant iron-iron contact area when particles collided. In the experiments, the inlet gas flow rate of N_2 was constant, implying that particles had essentially the same momentum. The sticking at 973 K was attributed to the adhesive force determined by the surface energy of metallic iron. Golunski[23] showed that Tammann temperature of metallic iron was 903 K, at which the solid point unit became noticeably mobile, resulting in markedly increasing the surface energy of metallic iron. Hence, high surface energy of metallic iron above Tammann temperature might be a dominant driving force for sticking. And the surface energy of metallic iron increased with temperature, therefore, the sticking time decreased with increasing temperature.

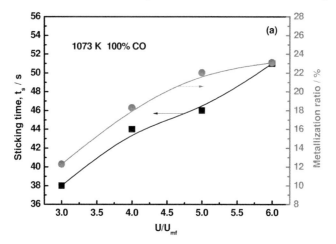

Figure 5   *Variations of sticking time and metallization ratio with gas velocity*

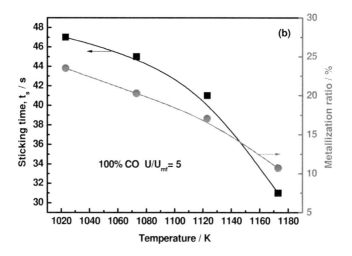

Figure 6   *Variations of sticking time and metallization ratio with temperature*

4 CONCLUSIONS

The pressure drop was examined during reduction of Fe_2O_3 particles with CO in a fluidized bed. Based on the pressure drop, the sticking time was introduced to describe the sticking characteristic. And the influence of fluidization conditions on the sticking time was investigated. The sticking time decreased with increasing CO concentration and reduction potential as a consequence of increasing iron-iron contact area when particles collided. However, the sticking time increased with gas velocity, resulting from increasing momentum of particles. Due to the increase of the surface energy of precipitated iron, as a dominating driving force for sticking, the sticking time decreased as the temperature increased.

Acknowledgement

This work is supported by National Natural Science Foundation of China and Baosteel under the grant 50834007.

References

1   M. Komatina, S. Oka, B. Grubor, and D. Voronjec, *Tenth International Heat Conference, Brighton*, 1994, **3**, 215.
2   M. Komatina, M. Ilic, and S. Oka, *14th International Conference on Fluidized Bed Combustion*, Vancouver, Canada, 1997, **2**, 863.
3   Habermann, F. Winter, H. Hofbauer and J. Zirngast, *ISIJ Int.*, 2000, **40**, 935.
4   F.J. Plaul, C. Bohm and J.L. Schenk, *J. S. Afr. Inst. Min. Metall.*, 2009, **108**, 121.
5   J.F. Fan, W.G. Li, Y.S. Zhou and Z.Y. Li, *Iron Steel*, 2007, **42**, 80.
6   X.Y. Hang and X.L. Zhou, *Mater. Metall. Eng.*, 2007, **35**, 49.
7   X.Y. Hang and X.L. Zhou, *Mater. Metall. Eng.*, 2008, **36**, 54.
8   S. Hayashi and S. Sawai and Y. Iguchi, *ISIJ Int.*, 1993, **33**, 1078.
9   M. Kimatina and H. W. Gudenau, *MJOM*, 2005, 309.
10   S. Hayashi and Y. Iguchi, *ISIJ Int.*, 1992, **32**, 962.
11   J. Fang, *Iron Steel*, 1991, **26**, 11.
12   S. Hayashi, S. Sayama and Y. Iguchi, *ISIJ Int.*, 1990, **30**, 722.
13   R. Degel, IEHK, RWTH, Aachen, 1996, 18.
14   F.M. Stephens and B.G. Langston, *J. Electrochem. Soc.*, 1960, **107**, C70.
15   Y.W. Zhong, X.Z. Gong, Z. Wang, Z.C. Guo, *J.USTB*, 2011, **33**, 406
16   B. Zhang, X.Z. Gong, Z. Wang, Z.C. Guo, *ISIJ Int.*, 2011, **51**, 1403.
17   A. Habermann, F. Winter, H. Hofbauer, J. Zirngast and J. L. Schenk, *ISIJ Int.*, 2000, **40,** 935.
18   S.P. Matthew and P.C. Hayes, *Metall. Trans. B*, 1990, **21B**, 153.
19   D.H. St. John. S.P. Matthew and P.C. Hayes, *Metall. Trans. B*, 1984, **15B**, 709.
20   R. Nicolle and A.Rist, *Metall. Trans. B*, 1979, **10B**, 429.
21   Z. S. Wu, R. T. Ma and Z. W. Wang, in *Fundamentals and Applications of Fluidized Bed Technology*, 1st Edn., Chemical Industry Press, Beijing, 2006, ch. 4, p. 69.
22   H.W. Gudenau, J. Fang, T. Hirata and U. Gebel, *Steel Res.* 1989, **60**, 138.
23   S. Golunski: *Platinum Met. Rev.*, 2007, **51**, 162.

SOLIDS FRICTION FACTOR FOR HORIZONTAL DENSE PHASE PNEUMATIC CONVEYING OF PULVERISED COAL

X. Guo, W. Li, H. Lu , X. Cong and X. Gong

Key Laboratory of Coal Gasification, Ministry of Education, East China University of Science and Technology, Shanghai, China
Email: gongxin@ecust.edu.cn

1   INTRODUCTION

Dense phase pneumatic conveying of pulverized coal is not only one of the core techniques for dry feed entrained-flow coal gasification technology, but also is extensively employed in chemical industry, metallurgical industry and power plant, etc. Predicting the pressure drop of pulverized coal for dense phase flow is often a key factor in designing a pneumatic conveying system to meet both technical and economical requirements.

As shown in Equation (1), Barth equation, based on power requirement,

$$\Delta P = \Delta P_g + \Delta P_s = (\lambda_g + \mu \lambda_z) \frac{L}{D} \frac{\rho_g U_g^2}{2} \tag{1}$$

has been employed extensively for pressure drop prediction due to its simplicity with accessible parameters. The main challenge is to determine the solids friction factor λ_z accurately. However, this equation was considered as applicable only to coarse particles in dilute phase flow,[1] and that numerous correlations for determining λ_z were based on experimental data of dilute pneumatic conveying. Molerus[2] also pointed out that the equation is not suited well for the pressure drop prediction of various particulate materials at rather low superficial gas velocities. However, it has been used subsequently to predict the pressure loss for the dense phase flow of fine powders by some researchers.[3,4]

Despite much progress has been made in the application of Barth equation for dense phase pneumatic conveying of powders, but so far the reported correlations for λ_z prediction can not give reasonable deviations for dense phase pneumatic conveying of pulverized coal and hardly meet the requirement of the entrained-flow gasification process of pulverized coal. In this paper, effects of pipe diameter, superficial gas velocity and solids volume fraction on the solids friction factor in horizontal pipelines were investigated, which results in a correlation for λ_z. Its accuracy for an industrial scale pipeline at high pressures was then examined.

2   METHODOLOGY

Figure 1 shows the schematic diagram of the experimental system. Pulverized coal in a feed hopper is transported by dry air into a receiving hopper with 3 gas streams injected into the feed hopper. Both the feed hopper and the receiving hopper have a volume capacity of 1.0 m³. The conveying pipeline is stainless steel and has flexible configuration in pipe diameter. Mass flow rate of pulverized coal is calculated by a weighing method. Static pressures down a horizontal pipeline were measured using 2 pressure sensors and the pipe pressure drop was obtained by calculating their difference.

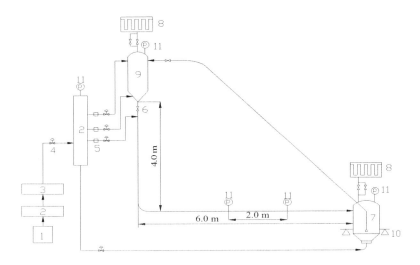

Figure 1 *Flow diagram of the experimental rig: 1. Air compressor; 2. Bunker; 3. Dryer; 4. Control valve; 5. Gas flow meter; 6. Ball valve; 7. Receiving hopper; 8. Filter; 9. Feed hopper; 10. Load cells; 11.Pressure sensor)*

Table 1   Key *experimental parameters*

Feed hopper volume (m³)		1.0
Receiving hoper volume (m³)		1.0
Pipeline inner diameter (mm)		15, 20, 26
Pipe length (m)	Vertical section	4
	Horizontal section	6

Table 1 details the test rig parameters. The total length of the pipeline is 10 m, consisting of 4 m downward vertical section and 6 m horizontal section with 15 mm, 20 mm and 26 mm inner diameter, respectively.

Gas friction factor λ_g due to gas alone can be calculated using Equation (2), which has been used with success for predicting gas friction loss.

$$\lambda_g = 0.3164 / \mathrm{Re}^{0.25} \qquad 2320 < \mathrm{Re} < 10^5 \tag{2}$$

Solids volume fraction (1-ε) can be estimated by

$$\lambda_z = \begin{cases} 0.005 + 0.63 Fr^{-1.89} & 4 < Fr \le 15 \\ 0.018 Fr^{-0.23} & 15 < Fr < 40 \end{cases} \tag{3}$$

3   RESULTS AND DISCUSSION

Figure 2 shows the pneumatic conveying profiles with a total of 481 individual measurement points for the three pipe diameters. It can be seen that, the solid/gas ratio ranges from 100 to 650 with the corresponding superficial gas velocity from 2 m/s to 15 m/s, which certainly falls into the dense phase pneumatic conveying. On the other hand, their mass flux ranges are identical from 1,000 kg/(m²·s) to 4,500 kg/(m²·s).

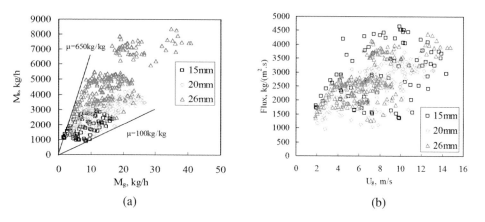

(a)   (b)

Figure 2   *Conveying profiles for different pipe diameters*

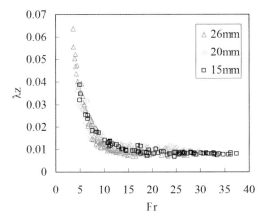

Figure 3 *Relationship between λ_z and Fr for different pipe diameters*

The solids friction factor λ_z is back calculated according to Equations (1) and (2). As shown in Figure 3, there is a good correlation between λ_z and Fr. Also, it can be easily seen that the pipe diameter has little effects on λ_z. Resultantly, a correlation for λ_z determination was obtained as shown in Equation (3) with Fr varying from 5 to 40. Figure 4 indicates the error distribution of calculated λ_z with Fr. It can be seen that the majority of deviation in λ_z calculation is generally within ±20%, which is reasonably acceptable for industrial application in the light of the complicated nature of the dense phase gas-solid flow.

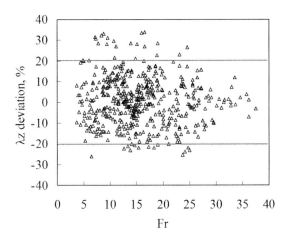

Figure 4 *Error distribution of λ_z calculated using Equation (3)*

Horizontal pneumatic conveying tests were carried out with 42 mm inner diameter pipe. Figure 5 shows the operation pressures with the mass flow rate ranging from 5 t/h to 15 t/h.

As a result, the pressure drop for the 42 mm pipe can be calculated using Equation (3). As shown in Figure 6, the pressure drop prediction is in a reasonable agreement with measured results. On the other hand, it suggests that the variation of pipe pressure has little effects on pressure loss predicted with Equation (3).

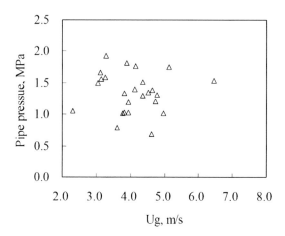

Figure 5   *Pipe pressure profile in pneumatic conveying of pulverized coal with I.D. 42 mm pipe*

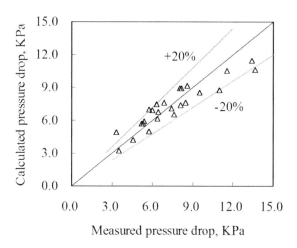

Figure 6   Calculated *pressure drop versus measured pressure drop*

4 CONCLUSIONS

A correlation of solids friction factor for the pressure drop prediction in horizontal dense phase pneumatic conveying of pulverized coal was obtained according to Barth Equation, from pneumatic conveying experiments of a pulverized coal with three pipe diameters at pressures of less than 0.4 MPa. It has been shown that the calculated pressure drops using the solids friction factor correlation are in good agreement with the measured ones in an industrial scale conveying plant, in which the tests were carried out with I.D. 42 mm pipes at high pressures up to about 2.0 MPa. It is also shown that the pipe pressure and diameter have little effects on pressure loss using this correlation.

Nomenclature

D	pipe diameter, m
ΔP	pressure drop, Pa
ΔP_g	pressure drop of gas phase, Pa
ΔP_s	pressure drop of solid phase, Pa
Fr	Froude number, $U_g/(gD)^{0.5}$
L	pipe length, m
M_g	mass flow rate of gas, kg/h
M_s	mass flow rate of solid, kg/h
Re	Reynolds number, $\rho ud /\mu$
U_g	superficial gas velocity, m/s
λ_g	Gas friction factor
ρ_g	density, kg/m^3
λ_z	solids friction factor
μ	mass flow rate ratio of solid and gas, M_s/M_g

Reference

1   M. Weber, *Bulk Solids Handl.*, 1981, **1**: 57.
2   O. Molerus, *Powder Technol.*, 1996, **88**: 309.
3   S. S. Mallick and P. W. Wypych, *Part. Sci. Technol.*, 2010, **28**: 51.
4   M. G. Jones and K. C. Williams, *Part. Sci. Technol.*, 2003, **21**: 45

DENSE PHASE PNEUMATIC CONVEYING OF BIOMASS AND BIOMASS/COAL BLENDS AT HIGH PRESSURES

C. He,[1] J. Wang,[1] H. Ni,[1] X. Chen,[1] Y. Xu,[1] H. Zhou,[1] Y. Xiong[1] and X. Shen [2]

[1] School of Energy and Environment, Southeast University, People's Republic of China
[2] School of Energy and Environment, Southeast University, People's Republic of China.
E-mail: xlshen@seu.edu.cn

1 INTRODUCTION

Due to global economic growth, the need for energy is increasing. As one of the most important fossil fuels in the world, coal plays a significant role. However, with the depletion of fossil fuel sources and the global warming issue, biomass, as a renewable energy source, becomes important. As a promising utilization method, biomass gasification attracts increasing attention. Pneumatic conveying is a key technology for both coal and biomass gasification. Hereinto, dense phase pneumatic conveying of powders is a competitive technology due to its high efficiency and low consumption of conveying gas and power.

Pneumatic conveying technology for solid materials has been developed over one hundred years and a large collection of valuable results have been accumulated in previous studies. Pan investigated the flow characteristics of pneumatic conveying of various solid materials, and pointed out that the flow mode for a bulk solid material is primarily dependent on material properties.[1] Geldart and Ling examined the transport of two sized fine coals using nitrogen, hydrogen, and carbon dioxide, they evaluated the individual contributions to the pressure loss caused by bends, acceleration and solids/wall friction and presented a correlation for pressure drops based on more than 600 data points, and they also investigated saltation velocities.[2,3] Southeast University performed pneumatic conveying experiments of pulverized coals with conveying pressure up to 4.0 MPa, and investigated the influences of coal properties and operation conditions. Many valuable insight into pneumatic conveying were obtained.[4-6] Wolfe et al. developed a pneumatic feeder for corn silage and haylage, experiments were performed to determine the relationship between pressure drop and conveying distance, air and material flow rates, and tube diameter for chopped forage.[7] Raheman and Jindal experimentally explored pneumatic conveying of rough rice, milled rice and soybeans and determined slip velocities for both horizontal and vertical pipe. They also examined the mixture characteristics.[8]

However, conveying biomass particles are challenging due to their unusual physical properties, particularly for dense phase pneumatic conveying at high pressures, and little work has been published. Therefore, intensive research on conveying characteristics of biomass is of great significance, and it offers a rich topic for research to improve transport properties of biomass and coal blends compared to biomass only. The objective of this present study was to explore experimentally the conveying characteristics in dense phase pneumatic conveying with three different materials, rice husk powder and two blends. The flow stability of these systems was also probed.

2 EXPERIMENTAL

Materials with a wide range of physical properties were considered. Tests were carried out using an independently developed pneumatic conveying system[4] under a variety of pressures and gas flows.

2.1 Materials

Rice husk powder (Material A) and two blends of the rice husk with a coal of different proportions: mass ratios of the coal to the rice husk was 1:2 (Material B) and 2:1 (Material C) , respectively, were used and their physical properties were summarized in Table 1.

Table 1   *Physical properties of the materials considered*

Material	Particle density (kg/m^3)	Mean particle size (μm)	Moisture (%)[a]
Material A	1016	67.81	9.55
Material B	1144	65.21	7.12
Material C	1272	62.61	4.89

[a] Total moisture content

2.2   The Pneumatic Conveying Equipment

The dense phase conveying system is comprised of a gas supply system, a material storage system, a conveying pipe loop, measurement sensors, data acquisition and control system, etc, as shown in Figure 1. The material storage system contains two vessels with a capacity of 0.648 m^3 each. Either vessel can be the sending or receiving vessel. The conveying pipeline is made of a smooth stainless steel tube with an internal diameter of 10 mm and a length of about 53 m. More details can be found in Shen *et al.*[4]

2.3   Operating Conditions

The influence of operating conditions on conveying characteristics were examined by i) varying the total conveying differential pressure through regulating the sending vessel pressure; and ii) changing the supplemental gas regulated with a gas flowmeter.

3 RESULTS AND DISCUSSION

There is a marked difference in conveying characteristics of the three materials tested. This is analysed in terms of solid mass flux, solid volume flux, superficial gas velocity and solid loading ratio.

Solid mass flux and solid volume flux are important characteristic parameters, and can be determined as follows:

$$G_{s,flux} = M_s / A \qquad\qquad (1)$$

$$V_{s,flux} = M_s / \rho_s / A \qquad\qquad (2)$$

where M_s is the mass flow rate through the pipeline measured via three load cells; ρ_s is the density of solids; A is the cross-sectional area of the pipeline.

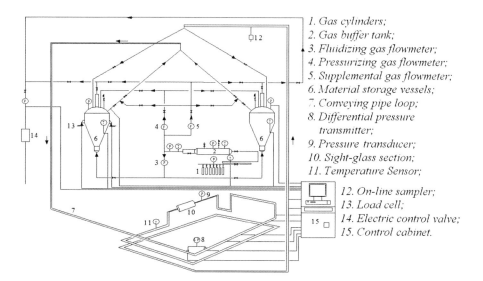

1. *Gas cylinders;*
2. *Gas buffer tank;*
3. *Fluidizing gas flowmeter;*
4. *Pressurizing gas flowmeter;*
5. *Supplemental gas flowmeter;*
6. *Material storage vessels;*
7. *Conveying pipe loop;*
8. *Differential pressure transmitter;*
9. *Pressure transducer;*
10. *Sight-glass section;*
11. *Temperature Sensor;*
12. *On-line sampler;*
13. *Load cell;*
14. *Electric control valve;*
15. *Control cabinet.*

Figure 1   *Schematic diagram of the dense phase pneumatic conveying system*

3.1   Influence of Total Conveying Differential Pressure

The total conveying differential pressure offers the primary dynamic source in a pneumatic conveying system, which converts the gas pressure potential energy into the kinetic energy and potential energy of solids flow in the pipe and overcomes the pressure loss of gas-solid two phase flow in conveying process. It determines the transmission capacity of the system.

Figure 2 shows the effects of the coal fraction in blends on solid conveying characteristics, and it also illustrates the influence of the total conveying differential pressure. It can be seen from Figure 2a that the solid mass flux increases with the increase of coal fraction in the blends at the same total conveying differential pressure, while the solid volume flux decreases as the coal fraction in the blends increases (Figure 2b). As expected, the solid mass flux and the solid volume flux increase with the increasing total conveying differential pressure.

It is also clear that the superficial gas velocity decreases as the coal fraction increases, but increases as the total conveying differential pressure increases (Figure 2c). As the coal fraction increases, the equivalent density of conveying material increases, and this leads to an increase of solid frictional pressure drop in horizontal pipe and vertical head loss. Obviously, the specific power consumption of solids will increase, but the gas kinetic energy will decrease correspondingly at the same total conveying differential pressure, so will the superficial gas velocity.

It can be seen that, at the same total conveying differential pressure, the solid loading ratio (SLR) increases with increasing coal fraction (Figure 2d), which can also be deduced from the change in solid mass flux and superficial gas velocity (Figures 2a & 2c). The SLR gradually increases when the total conveying differential pressure is increased (Figure 2d).

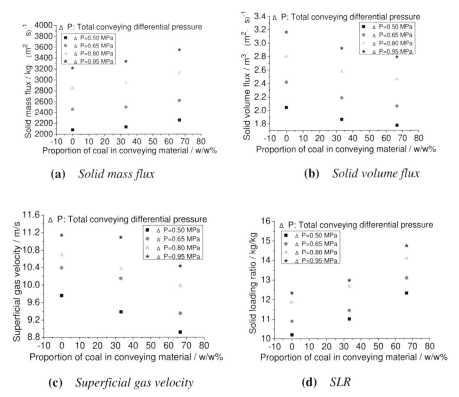

(a) *Solid mass flux*

(b) *Solid volume flux*

(c) *Superficial gas velocity*

(d) *SLR*

Figure 2 *Effects of coal fraction in the blends on conveying characteristics at different total conveying differential pressures (Δp)*

3.2   Influence of supplemental gas

The supplemental gas is one of the effective measures to ensure the smooth operation of a conveying system. How the supplemental gas affects conveying behaviours is also examined in detail and the results are presented in Figure 3.

It is evident from Figure 3a that the solid mass flux gradually increases with the increase of coal fraction, but decreases with increasing supplemental gas. The increase of supplemental gas can reduce the differential pressure between sending vessel and its outlet pipeline, causing the decrease of solid mass flux.

Figure 3b shows the relationship between solid volume flux and supplemental gas. It eliminates the influence of solid density that is involved in Figure 3a. It can be seen from Figure 3b that the solid volume flux decreases with increasing coal fraction. This is due to the fact that the rice husk powder has better flow performance and can be carried into the conveying pipeline more easily. It is clear from Figure 3c that the superficial gas velocity decreases as the coal fraction increases, but increases with increasing supplemental gas. SLR decreases with increasing supplemental gas as shown in Figure 3d. It is believed that, as the supplemental gas increases, the solid mass flow rate decreases, the contribution of the gas in two-phase flow increases, while that of the powder decreases, resulting in a smaller SLR.

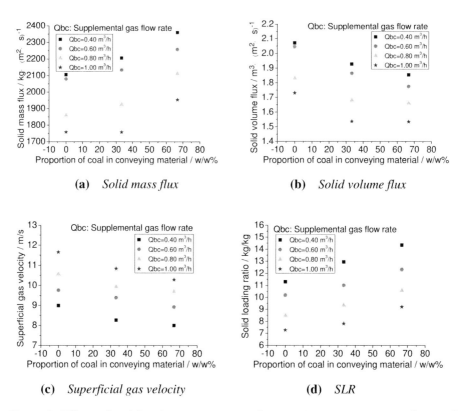

(a) *Solid mass flux*   **(b)** *Solid volume flux*

(c) *Superficial gas velocity*   **(d)** *SLR*

Figure 3 *Effects of coal fraction on conveying characteristics at various supplemental gas flow rates*

3.3   Flow stability

Stability of the feeding system is an essential operation condition for gasification, and it can be approximately reflected by flow stability in pneumatic conveying. Relative standard deviation (RSD) analysis is an effective method for stability analysis. Figures 4 and 5 illustrate how the RSD of powder mass flow rate varies with the total conveying differential pressure and the supplemental gas. The RSDs are all maintained at 4% or less. Thus, it can be concluded that the pneumatic conveying system has a good stability in this study. Note that, the RSD for conveying two blends is significantly smaller than that for conveying rice husk alone. It is expected that the flow stability of blends was better than the pure rice husk. It should also be noted that the plugging problem occurring at high vessel pressures for conveying rice husk was avoided in conveying the blend. It is anticipated that the increased friction between highly compressed rice husk powder particles and pipe wall leads to the reduction in particle velocity and plugging. With the addition of coals, the friction is reduced and the plugging problem is hence avoided.

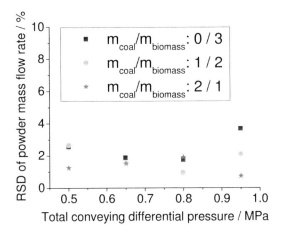

Figure 4 *Variation of RSD of powder mass flow rate with the total conveying differential pressure*

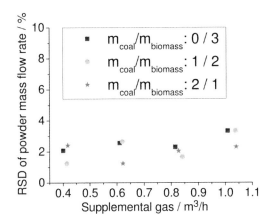

Figure 5 *Variation of RSD of powder mass flow rate with supplemental gas*

4 CONCLUSIONS

Pneumatic conveying experiments were carried out for three different materials at high pressures. The effect of operation conditions on conveying behaviours was also investigated. Distinct differences in conveying characteristics of different materials were observed. Effects of material properties and operational parameters on conveying behaviours were analysed in terms of solid mass flux, solid volume flux, superficial gas velocity and solid loading ratio. It was shown that under the same operation conditions, as the coal fraction increases, solid mass flux and solid loading ratio increase while solid volume flux and superficial gas velocity decrease gradually. Transport properties of biomass/coal blends were greatly improved compared to rice husk only. The flow stability of blends was better and the phenomenon of instability and plugging during conveying rice husk powders at high solid loading ratios was avoided. It was found that with the increase of total conveying differential pressure, solid mass flux, solid volume flux, solid loading ratio and superficial gas velocity increase. Conversely, with the increase of supplemental gas, solid mass flux, solid volume flux and solid loading ratio decrease, while the superficial gas velocity increases.

Finally, it should be pointed out that, in addition to the conveying characteristics concerned in this study, the pressure drops in the horizontal pipe, vertical pipe, horizontal bend and vertical bend are also important characteristic parameters and depend on mixing proportions of powder blends, which deserve further investigation.

Acknowledgements

Financial support for this work was provided by the National Basic Research Program of China (973 Program) (2010CB227002).

References

1   R. Pan, *Powder Technol.*, 1999, **104**, 157.
2   D. Geldart and S. J. Ling, *Powder Technol.*, 1990, **62**, 243.
3   D. Geldart and S. J. Ling, *Powder Technol.*, 1992, **69**, 157.
4   X. Shen and Y. Xiong, *Proc. CSEE*, 2005, **25**, 103. (In Chinese)
5   C. Xiaoping, F. Chunlei, L. Cai, P. Wenhao, L. Peng and Z. Changsui, *Korean J. Chem. Eng.*, 2007, **24**, 499.
6   X. Pan, C. Xiaoping, L. Cai, Z. Changsui, L. Bo, X. Guiling and L. Haixin, *Journal of China Coal Society*, 2010, **35**, 1359. (In Chinese)
7   R. R. Wolfe, M. M. Smetana and G. W. Krutz, *Trans. ASAE*, 1970, **339**, 332.
8   H. Raheman, and V.K. Jindal, *Powder Handl. Process.*, 1994, **6**, 29.

Modelling

THE INFLUENCE OF WETTING ON THE BUOYANCY OF PARTICLES

J. Bowen,[1] D. Cheneler,[2] M.C.L. Ward[2] and M.J. Adams[1]

[1]School of Chemical Engineering, The University of Birmingham, Edgbaston, Birmingham, B15 2TT, UK
[2]School of Mechanical Engineering, The University of Birmingham, Edgbaston, Birmingham, B15 2TT, UK

1 INTRODUCTION

The role of wetting in particulate systems can be of practical importance and has been considered for many decades, with Huh and Mason[1] providing an early example of a rigorous theoretical consideration of a single particle buoyant on the surface of a liquid. The ability of a particle of a density greater than the liquid phase on which it rests to be buoyant will be significantly influenced by the wetting interaction that occurs at the interface between the solid and liquid phases. Recent work by Vella *et al.*[2] and that of Extrand and Moon[3-5] has sought to capture the importance of wetting on the buoyancy of particles.

In this work, the effect of wetting on the ability of solid millimetre-sized and smaller particles to float on a liquid of lower density than the particle material has been considered theoretically and subsequently investigated experimentally. The contribution of capillary force towards particle buoyancy is assessed for a range of systems, as well as the range of diameters of particles which could be buoyant for each combination of solid and liquid. The critical parameters in this evaluation of particle buoyancy were considered to be (i) solid density, (ii) liquid density, (iii) liquid surface tension, and (iv) interfacial wetting between solid and liquid phases.

2 THEORETICAL

A force balance was performed for a single solid particle in a semi-infinite volume of fluid; the capillary (F_C), buoyancy (F_B) and gravitational forces (F_G) were considered. Figure 1 shows the system force balance. The capillary force includes contributions from both surface tension and Laplace pressure. The Laplace pressure was calculated from the Young-Laplace equation by assuming the meridional profile is approximated by a circular arc, as derived by Orr *et al.*[6] Buoyancy forces and gravitational forces were calculated by assuming a spherical geometry for the particle. The force balance provides a criterion for whether a given particle will be buoyant. Figure 2 shows the system geometry, in which the particle radius, R, liquid/solid wetting angle, θ, half-filling angle, ψ, submersion depth, D, and height of the exposed spherical cap above the liquid surface, h, were all considered.

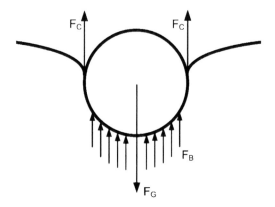

Figure 1 *System force balance for a solid particle buoyant on the surface of a liquid*

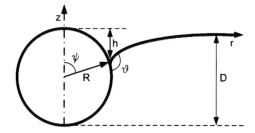

Figure 2 *System geometry balance for a solid particle buoyant on the surface of a liquid*

The gravitational force, F_G, is described by Eqs. 1-3, where m_P is the particle mass, g is the acceleration due to gravity, ρ_P is the particle density, and V_P is the particle volume.

$$F_G = -m_P g \tag{1}$$

$$F_G = -\rho_P g V_P \tag{2}$$

$$F_G = -\frac{4}{3}\pi \rho_P g R^3 \tag{3}$$

The buoyancy force, F_B, is described by Eqs.4-7, where m_L is the mass of displaced liquid, ρ_L is the liquid density, V_L is the volume of displaced liquid, which is assumed to be equivalent to the particle volume minus the volume of a spherical cap, the diameter of which is defined by the half-filling angle, ψ.

$$F_B = m_L g \tag{4}$$

$$F_B = \rho_L g V_L \tag{5}$$

$$F_B = \rho_L g \left[\frac{4}{3}\pi R^3 - \frac{1}{3}\pi h^2 (3R - h) \right] \tag{6}$$

$$\text{where, } h = R[1 - \sin(\psi)] \tag{7}$$

The capillary force, F_C, is described by Eqs. 8-10, where γ_L is the liquid surface tension, and H is the dimensionless mean curvature of the liquid surface, as described by Orr *et al.*,[6] which is assumed to be a circular arc.

$$H = \frac{\sin(\varphi+\psi)}{R\sin(\psi)} \tag{8}$$

$$F_C = 2\pi\gamma_L R[\sin(\psi)\sin(\varphi + \psi) - HR\sin^2(\psi)] \tag{9}$$

$$F_C = 4\pi\gamma_L R[\sin(\psi)\sin(\varphi + \psi)] \tag{10}$$

The system force balance is shown in Eqs. 11-12, in which F_T is the total force. A particle is considered buoyant if $F_T > 0$.

$$F_T = F_B + F_C + F_G \tag{11}$$

$$F_T = \rho_L g \left[\frac{4}{3}\pi R^3 - \frac{1}{3}\pi h^2 (3R - h) \right]$$
$$+ 4\pi\gamma_L R[\sin(\psi)\sin(\varphi + \psi)] - \frac{4}{3}\pi\rho_P g R^3 \tag{12}$$

Table 1 *Particle and liquid densities measured at 20 °C using helium pycnometry*

Material	Density / kg m^{-3}
Aluminium oxide	4,000
Glass	2,200
Glycerol	1,261
Nylon	1,150
Poly(tetrafluoroethylene)	2,200
Poly(urethane)	1,050
Water	1,000

3 EXPERIMENTAL

Particle buoyancy was assessed by placing a particle of interest into a clean glass vessel, followed by the gradual addition of the liquid of interest around the particle, until the height of the liquid surface was significantly greater than the original height of the top of the particle. The particles employed in this study were aluminium oxide, glass, nylon, poly(tetrafluoroethylene), and poly(urethane), of diameters in the range 1-10 mm, all of which were purchased from Dejay Distribution (UK). The liquids employed in this study were HPLC grade water (Sigma-Aldrich, UK) and standard laboratory grade glycerol (Fisher Scientific, UK). Particle and liquid densities were measured using an AccuPyc II 1340 helium pycnometer (Micromeritics, UK). Densities for all materials used in this study are listed in Table 1. Particle surface roughnesses were measured using a MicroXAM2 white light interferometer (Omniscan, UK). Particle surface roughnesses were < 1 % of the reported particle diameters. The wetting angles, or contact angles, of the liquids on the solid particles was measured using a home-made apparatus.

4 RESULTS AND DISCUSSION

Figure 3 shows a poly(tetrafluoroethylene) particle of diameter 3.96 mm floating in a bath of water. It can be seen that there is a curvature to the liquid surface, which is at a maximum close the particle/liquid interface, and decreases further away from the interface. The surface curvature is essentially zero at a radial distance of approximately 8 mm away from the particle centre. Further work is currently underway to investigate the radial extent and three-dimensional geometry of this curvature for a range of liquid/particle combinations, and attempt to relate this to the Young-Laplace equation.

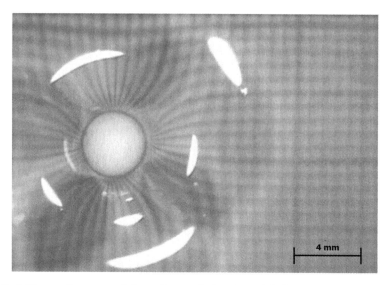

Figure 3 *3.96 mm diameter poly(tetrafluoroethylene) particle buoyant on a water surface*

Figure 4 shows the calculated result for a 3.96 mm poly(tetrafluoroethylene) particle on a water surface. F_T is shown on the z-axis, as a function of half-filling angle (x-axis) and contact angle (y-axis). The calculation shows the range of permissible solutions for a given particle/liquid combination, and the solutions permissible for poly(tetrafluoroethylene)/water can be identified at the measured contact angle, 105 °. Where $F_T > 0$, there exist a range of half-filling angles at which the particle should be buoyant. Hence, the model predicts that a 3.96 mm diameter poly(tetrafluoroethylene) particle will be buoyant on a water surface, a result which is in agreement with the experimentally assessed result shown in Figure 3. At a half-filling angle of 50 °, $F_C = 1.05$ mN, $F_B = 0.25$ mN, $F_G = -0.7$ mN, and $F_T = 0.6$ mN, a result which indicates that capillary forces can provide a significant contribution to particle buoyancy when the liquid phase exhibits a high wetting angle on the solid particle surface.

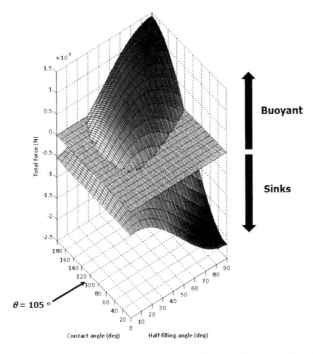

Figure 4 *Calculated result for a 3.96 mm diameter poly(tetrafluoroethylene) particle on a water surface*

Table 2 *Particle/liquid combinations assessed for buoyancy during this study*

Particle	Liquid	Wetting angle/$^\circ$	$2R_{max}$/mm	ψ at R_{max}/$^\circ$
Poly(urethane)	Water[a]	85	10.0	58
Poly(tetrafluoroethylene)	Water	105	6.3	50
Poly(tetrafluoroethylene)	Glycerol[b]	85	3.2	52
Nylon	Water	85	2.8	45
Aluminium oxide	Water	90	1.9	45
Glass	Water	0	n/a	n/a

[a] γ_{water} = 72 mN/m at 20 $^\circ$C. [b] $\gamma_{glycerol}$ = 64 mN/m at 20 $^\circ$C.

Table 2 summarises the calculated results that have been experimentally verified during this study. Note that in all cases considered, the particle density is greater than the liquid density. It can be seen that the poly(urethane) particles of diameters up to 10 mm are buoyant on a water surface, whereas poly(tetrafluoroethylene) particles of diameters up to 6.3 mm are buoyant on a water surface. poly(urethane) has a lower density than poly(tetrafluoroethylene), but the wetting angle of water on a poly(tetrafluoroethylene) surface is greater than that of water on a poly(urethane) surface. Buoyancy cannot be simply assessed without a consideration of the material densities and also the particle/liquid wetting properties. Table 2 shows that the maximum buoyant particle diameter decreases with increasing particle density. The filling angle at the maximum particle radius is in the range 45-60 °, and this will be the subject of further study, aiming to correlate theoretical predictions with experimental measurements. In Figure 4 it can also be seen that there are no buoyant solutions where the wetting angle is < 50 °, and this result was observed to be generally true regardless of particle diameter, unless the density of the particle was similar to that of the liquid phase.

5 CONCLUSION

A theoretical model has been developed in which the particle/liquid wetting properties have been incorporated into an assessment of particle buoyancy. The maximum permissible particle diameter has been calculated for a range of particles and liquids, exhibiting different densities and particle/liquid wetting angles. The theoretical results are in good agreement with experimentally verified assessments of buoyancy. The model reveals that capillary forces provide a significant contribution to particle buoyancy when the liquid phase exhibits a high wetting angle on the solid particle surface. The model also revealed that particles exhibiting wetting angles less than 50 ° when in contact with the liquid phase tended not to be buoyant, regardless of their diameter, unless the density of the particle was similar to that of the liquid phase.

Acknowledgements

The University of Birmingham and Unilever Research & Development are acknowledged for financial support for JB and DC. The Interferometer and Helium Pycnometer used in this research was obtained, through Birmingham Science City: Innovative Uses for Advanced Materials in the Modern World (West Midlands Centre for Advanced Materials Project 2), with support from Advantage West Midlands (AWM) and part funded by the European Regional Development Fund (ERDF).

References

1. C. Huh and S.G. Mason, *J. Colloid Interface Sci.*, 1974, *47*, 271-289.
2. D. Vella, D-G. Lee, H-Y. Kim, *Langmuir*, 2006, *22*, 5979-5981.
3. C.W. Extrand and Moon, S.I., *Langmuir*, 2009, *25*, 992-996.
4. C.W. Extrand and Moon, S.I., *Langmuir*, 2009, *25*, 2865-2868.
5. C.W. Extrand and Moon, S.I., *Langmuir*, 2009, *25*, 6239-6244.
6. F.M. Orr, L.E. Scriven and A.P. Rivas, *J. Fluid Mech.*, 1975, *67*, 723-742.

TRANSPORT PHENOMENA IN PACKED BEDS

R. Ocone

Chemical Engineering, Heriot-Watt University, Edinburgh EH14 4AS, UK

1 INTRODUCTION

Most of the energy in the world derives from combustion of fossil fuel. One of the major environmental concerns is that fossil fuel combustion releases CO_2, which is believed to be linked to global warming and climate change. Given that the usage of fossil fuel is not expected to decrease in the future, governments are pushing for a major change to diversified energy supply and, possibly, to a more sustainable usage of fossil fuel with less environmental impact. Chemical looping combustion (CLC) is potentially the technology best suited for capturing CO_2 at low cost and efficiently providing a low energy option for the separation of CO_2 from flue gases. The process consists in cyclic reduction and oxidation of a metal which acts as oxygen carrier and which is exchanged between two reactors, usually a circulating fluidised bed (the oxidation reactor) and a bubbling bed reactor (the combustor). The oxygen carrier transfers oxygen from the air to the fuel, hence a direct contact between air and fuel is avoided. Consequently, the outlet gas from the combustor contains only CO_2 and H_2O: the latter is easily removed by condensation and the CO_2 is readily captured for storage and/or utilisation.[1] The main attraction of the process is that it does not involve any penalty for extra highly intensive separation of CO_2 from the other combustion products.

Noorman et al.[2,3] introduced a new concept to carry out the oxidation-reduction cycle, whilst still retaining the concept of utilising metal particles as oxygen carrier: instead of considering two reactors, they use a packed bed reactor which is dynamically operated where the gas flow is switched and the bed of particles is stationary, alternating an oxidation cycle to a reduction (combustion) cycle. There are inherent advantages in operating a fixed bed instead than fluidised beds; among other drawbacks, the condition in the riser (high velocity) and those in the bubbling bed (low velocity) are often incompatible to generate good particle distribution, gas-solid contact and adequate residence time to achieve the conversion and the thermal output needed. Downers seem to offer advantages over the most favourite set-up and less sensitive to operational changes; however, the solid hold-up and heat transfer might be a limitation. To obtain a reasonable thermal output, the kinetics and thermodynamics of the process require a high circulation rate of solids. This means that the top-to-bottom mixing of the solids in each reactor must be small. In addition, the metallic oxide has a high particle density and it is not so clear whether existing correlations for equilibrium bubble size, bed expansion and particle entrainment are applicable; the particle density/particle size envelope for Group A powders becomes smaller for high density materials, that is, the choice of the mean particle sizes available, while still remaining in Group A, is much smaller than with Fluid Catalytic Crackers, for example. Finally, the process is associated to inevitable particle attrition with the consequent need of very efficient separation after the riser to avoid damages from fines in the hot air to the gas turbine.[4]

The works of Noorman and co-workers[2,3] have shown the feasibility of the concept of packed bed and their experiments have shed light on the limits of operability of the process. They have concluded that the high temperature air stream can be produced and sustained in this reactor configuration and, since the maximum temperature is independent of the gas mass flow rate, the set-up shows high operating flexibility. The findings, based on the assumption that the kinetics of the oxidation cycle is extremely fast, are an important starting point for developing the analysis presented in this paper. In relation to the classification given by Ocone and Astarita,[5] this is discontinuous model for the mass transport phenomenon, where the change from un-reacted to reacted solids is abrupt, with a sharp front which moves through the solid bed.

2 BALANCE EQUATIONS

In this section, the methodology which is known as the Kotchine Theorem (KT) is briefly introduced and applied to a one-dimensional problem, namely a packed bed of particles of length, L. The system can be considered one-dimensional, if no radial dispersion is considered and variations are then assumed to happen only along the axial position, x. If r is the rate of supply of a given quantity c, and q is its flux, the balance for unit volume for that variable c is given by:

$$\frac{\partial q}{\partial x} + \alpha \frac{\partial c}{\partial t} = r \qquad (1)$$

where α is a constant expressing the physical characteristics of the system (e.g., specific heat, density, gas velocity, etc.) that will make the equation dimensionally correct (depending on the quantity for which the balance is written). c, q and r are in general function of the spatial variable, x, and time, t.

To close the problem, constitutive equations for q and r need to be supplied together with the required boundary conditions. Before attempting the formal solution of Eq. (1), it is worth noting that, if a discontinuity is assumed to travel across the bed, then, at the two sides of the discontinuity, the characteristics of the bed and the values of the relevant variables are expected to be different. Therefore the problem would be amenable to being treated as a moving boundary problem. A balance equation on the surface of the discontinuity can be obtained by writing the balance on the moving surface or following a more formal procedure known as the KT procedure.[4]

In the following, to illustrate the methodology, we refer to Figure 1 and let $[c]$ be the jump of the variable c across the discontinuity, i.e. $[c] = c_R - c_L$, where c_R is the value of c on the right of the discontinuity and c_L its value on the left.

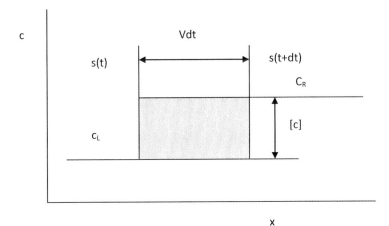

Figure 1 *Direct route for obtaining balance equations across discontinuities*

During the time interval dt, the discontinuity has moved over a distance Vdt. Correspondingly, an amount $V[c]dt$ (per unit surface) has been removed from the dark area in Figure 1. If r is finite, the rate of generation per unit surface area is zero, and the flux is responsible for the removal of the variation of c; the amount removed is $[q]dt$, and therefore the balance furnishes:

$$[q] = V[c] \qquad (2)$$

Suppose now that the discontinuity can be located at each time by the relation $y = s(\tau)$, with $V = \dfrac{ds}{d\tau}$ being the speed of propagation of the discontinuity within the medium. By defining the new variable $y = (s - x)$, the relationships follow:

$$\frac{\partial}{\partial x} = -\frac{\delta}{\delta y}$$

$$\frac{\partial}{\partial t} = \frac{\delta}{\delta \tau} + V\frac{\delta}{\delta y} \qquad (3)$$

where δ indicates the partial derivative in the y-t system of independent variables. Both q and c are differentiable on the right and left of the discontinuity and by taking the integral of Eq. (1) between $y = -\varepsilon$ and $y = +\varepsilon$, and the limit when $\varepsilon \to 0$, then Eq. (2) is obtained. This procedure of changing variables and integrating the balance on the discontinuity is called Kotchine theorem.[4] If there is a finite rate of supply per unit surface, the balance equation becomes:

$$[q] = V[c] + r' \qquad (4)$$

which represents the generalised KT procedure.[6,7] In most of the literature, the KT was applied to balance equations until Ocone and Astarita[4] applied the procedure to constitutive equations, showing both the flexibility and the power of the procedure in determining characteristic velocities when shocks and waves develop in the problem under scrutiny. To obtain the mass front (reaction) velocity, V, the continuity equation (e.g. Eq. (1) applied to mass conservation) must be written. Analogously, the energy balance should furnish the heat front velocity.

Astarita and Ocone[8] investigated the predictive nature of the KT by applying it to various transport problems in the area of heat and mass transfer. Although they solved a similar problem, i.e., the combustion of a slab of coal, the KT has not been tested yet in the heat and mass transfer in a packed bed. In the following, we apply the procedure to the packed bed reactor used for the oxidation and reduction reactions in CLC technology as introduced by Noorman et al.[2] It will be shown that the mass front velocity can be obtained without explicitly solving the transport equations. Analogously, if the KT is applied to the energy balance, then the heat front velocity can be obtained.

3 MODEL FORMULATION

The mass balance of the oxygen diffusing in the packed bed of metal particles reads:

$$\frac{\partial c}{\partial t} = -v\frac{\partial c}{\partial x} + D\frac{\partial^2 c}{\partial x^2} + \dot{R} \tag{5}$$

where c is the concentration of oxygen, v its velocity, D its diffusivity and $\dot{R}$ is the rate disappearance of oxygen, namely the rate at which the oxygen is consumed by reacting with the solid.

We consider that the reaction only happens on the exposed surface of the solid particles; this is not an unrealistic assumption for some metal particles; additionally, and especially in packed beds, operating conditions (e.g., gas flow rate) can be controlled in a way that assures such assumption. Based on experimental evidence, the phenomenon resembles an adsorption (Γ) process and the oxidation is considered to be a surface phenomenon, namely surface oxides are formed. Formally, the reaction rate can be written as:

$$\dot{R} = -\frac{\partial \Gamma}{\partial t} \tag{6}$$

The adsorption can have different forms depending on the bed; however, if the kinetics of oxidation is known, then the dependence of Γ on the concentration c is known too. With such a substitution, Eq. (5) can be solved obtaining profiles of concentrations and the velocity of the mass front.

Let us assume now that the reaction is instantaneous (therefore at equilibrium); the following relation holds:

$$\frac{\partial \Gamma}{\partial t} = -\left(\frac{\partial \Gamma}{\partial c}\right)\left(\frac{\partial c}{\partial t}\right) \tag{7}$$

And substituting this expression into Eq. (5), we obtain:

$$\frac{\partial c}{\partial t} = -v\frac{\partial c}{\partial x} + D\frac{\partial^2 c}{\partial x^2} - \left(\frac{\partial \Gamma}{\partial c}\right)\left(\frac{\partial c}{\partial t}\right) \tag{8}$$

This is the general equation governing the problem and, once the explicit relation for $\Gamma(c)$ is assigned, then it can be solved to obtain the velocity profile of oxygen within the bed and its velocity of propagation. Eq. (8) can be rearranged to the following form:

$$\left(1 + \frac{\partial \Gamma}{\partial c}\right)\frac{\partial c}{\partial t} = -v\frac{\partial c}{\partial x} + D\frac{\partial^2 c}{\partial x^2} \tag{9}$$

or, equivalently:

$$\frac{\partial c}{\partial t} = -v'\frac{\partial c}{\partial x} + D'\frac{\partial^2 c}{\partial x^2} \tag{10}$$

where:

$$v' = \frac{v}{1 + \dfrac{\partial \Gamma}{\partial c}}$$

$$D' = \frac{D}{1 + \dfrac{\partial \Gamma}{\partial c}} \tag{11}$$

Consequently, the oxygen will now flow with a rescaled velocity v' and its diffusion is rescaled by the same factor, giving an effective diffusivity D'. This result is obtained without any assumption being made on the specific relation $\Gamma(c)$.

Eq. (10), when an explicit form for $\Gamma(c)$ is assigned, describes the problem fully. Let us assume that the dependence of Γ on c is linear, then $\dfrac{\partial \Gamma}{\partial c}$ is constant and the gas flows at constant velocity. In this particular case Eq. (10) is the classical diffusion-convection equation with modified values of flow rate and diffusivity.

4 THE KINETIC LIMIT

In this section we analyse the case where the diffusion is negligible. In this limit the (mass) transport equation for the oxygen becomes:

$$\frac{\partial c}{\partial t} = -v\frac{\partial c}{\partial x} - \frac{\partial \Gamma}{\partial t} \qquad (12)$$

Eq. (12) is hyperbolic and admits discontinuities; applying the KT to Eq. (12), we obtain:

$$V[c] = v[c] - V[\Gamma] \qquad (13)$$

This is a general result:

$$V = \frac{v[c]}{[c] + [\Gamma]} \qquad (14)$$

stating that the mass front depends on the jump of the adsorption at the front. The value of $[\Gamma]$ can be obtained by integrating Γ across the discontinuity or, equivalently, applying the KT to the function $\Gamma(c)$. However, to do so, a specific form for Γ needs to be assigned. Since $\Gamma(c)$ is in general non linear in c, its integration around the front can be quite tricky. However, Eq. (12) represents the governing transport equation and therefore the front can be found by solving it.

To show the applicability of the KT, without any need to solve the full mass balance, a simple expression for $\Gamma(c)$ is chosen: the function is approximated by a bi-linear equation where, below a critical concentration, c^*, the isotherm is given by $\Gamma = k\,c$ and above c^*, $\Gamma = \Gamma^*$, representing the maximum value of Γ (the maximum adsorption level has been reached and from this point on, the oxygen does not react with metal surface). This situation is very similar to a "stripping" isotherm, in this particular case representing the stripping of oxygen from the front.

Eq. (11), when the $\Gamma(c)$ is bi-linear, furnishes the following values for the gas velocity:

$$v' = \frac{v}{1+k} \qquad \text{if } c < c^* \qquad (15)$$

$$v' = v \qquad \text{if } c > c^* \qquad (16)$$

In this case, the diffusion phenomenon has different characteristics depending whether the maximum value of Γ, Γ^*, has been reached. Figure 2 shows how the concentration changes along the bed length. The assumption in Figure 2 is that, when the reaction is considered, $\frac{\partial \Gamma}{\partial c}$ is constant.

The dotted line represents the situation where c^* is taken equal to 0.2 and it is an interesting case to analyse. Up to c^*, the reaction takes place; the removal of oxygen from the gas stream means that the condition at the interface between reacted and un-reacted bed changes and that reflects in "slowing down" the interface that would be expected if no reaction is present: the solid black line in Figure 2 represents the solution for the convection-diffusion problem, where no reaction takes place.

The influence of the reaction is shown in Figure 3, where the concentration distributions, corresponding to three increasing values of the rate constant, k, are reported.

In addition, in Figure 3 it is assumed that c<c* everywhere in the bed, then the reaction is the only predominant phenomenon. It is evident that, as the rate of reaction increases, the front delay increases. The velocity of the front is predicted correctly from the KT, since the jumps of the relevant variables are known on the two sides of the front (reacted and un-reacted bed).

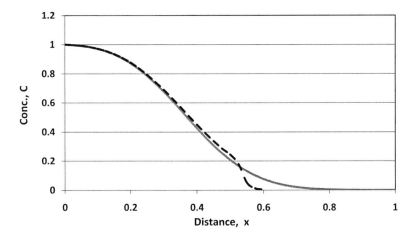

Figure 2 *Concentration along the bed when a stripping isotherm is assumed with c*=0.2 (---) and for the convection problem only (reaction and diffusion both null)*

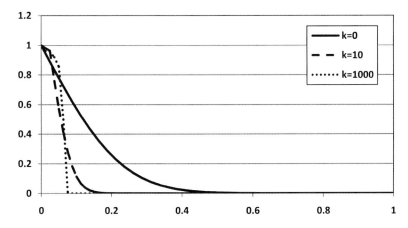

Figure 3 *Oxygen concentration along the bed when a stripping isotherm is assumed with c*=1 for increasing rate of reaction*

5 CONCLUSIONS

The treatment presented in the previous sections focuses on mass transfer (oxidation) considering the reaction completing and reaching equilibrium rapidly; this also implies that the reaction is a surface phenomenon, where the existence of two zones (the reacted and un-reacted ones) is assumed a priori. Even if this is a good approximation for the oxidation of metal particles, one cannot neglect to consider that the rate of evolution of the reaction can be finite; of course, when the intrinsic kinetic time scale becomes sufficiently small, the solution of the bulk kinetics should asymptotically approach the that of a surface kinetics, where the rate of reaction per unit surface is finite. The rate of reaction can however be incorporated in the analysis.

The mass transfer consequent to the oxidation of the oxygen-carrier particles has been considered. In chemical looping technology, the heat transfer problem must be solved. The main aim of this work is to show how the KT can furnish interesting insights into the evolution of the process. Not only the KT can be applied to the reduction (combustion) cycle, but it can be applied to heat transfer too.

A final point that is of importance is that the KT is extremely useful when a perturbation happens in the bed such as, for instance, a finite (or infinitesimal) disturbance can propagate in the flowing gas. The KT around the disturbance can furnish indication of whether the disturbance will reinforce or will die out, giving useful information for the operation of the process.

Acknowledgements

The author gratefully acknowledges the Royal Society of Edinburgh and the Scottish Government for supporting the present work under the Support Fellowship scheme. Useful discussion with Professor L. Nero is also acknowledged.

References

1 M.H. Hossain and H.I. de Lasa, *Chem. Eng. Sci.*, 2008, **63,** 4433.
2 S. Noorman, M. van Sint Annaland and J.A.M. Kuipers, *Ind. Eng. Res.,* 2007, **46**, 4212.
3 S. Noorman, M. van Sint Annaland and J.A.M. Kuipers, *Chem. Eng. Sci.*, 2010, **65**, 92.
4 R. Ocone, *IFP Energies Nouvelles*, Lyon, France, March 2010.
5 R. Ocone and G. Astarita, *AIChE Journal*, 1987, **33**, 423.
6 F.B. Foraboschi, *Principi di Ingegneria Chimica,* UTET, Torino, 1974.
7 G. Astarita and R. Ocone, in *Advances in Transport Processes*, A.S. Mujumdar and R.A. Mashelkar Eds., Elsevier, Amsterdam 1992, 319.
8 G. Astarita and R. Ocone, *Special Topics in Transport Phenomena*, Elsevier, Amsterdam, 2002.

INCORPORATING THE WEDGING EFFECT IN A NEW PACKING MODEL

A.K.H. Kwan, K.W. Chan and V. Wong

Department of Civil Engineering, The University of Hong Kong,
Pokfulam Road, Hong Kong, China. E-mail: khkwan@hku.hk

1   INTRODUCTION

The prediction of packing density of a granular system is of major importance in many branches of science and engineering. For the prediction of packing density, a number of packing models have been developed.[1,2] However, the theoretically predicted maximum packing density of a binary mix of mono-sized particles is often higher than that obtained by measurement. As the existing packing models have already taken into account the various known effects, such as the filling and loosening effects of the finer particles and the occupying and wall effects of the coarser particles,[3,4,5] the significant discrepancies between the predicted and measured results cannot be explained by any of the known effects. It is postulated herein that such discrepancies are caused by an effect called the wedging effect and that with this effect incorporated into a new 3-parameter packing model, we can better understand the packing structure of particles and improve the accuracy of our theoretical predictions.

2   DERIVATION OF THE 3-PARAMETER PACKING MODEL

2.1   The Loosening Effect Parameter a and the Wall Effect Parameter b

Let us consider a binary mix of mono-sized particles. Two scenarios may occur. When the coarse particles are dominant, the fine particles would fill into the voids between the coarse particles, thus increasing the packing density by means of *the filling effect*. On the other hand, when the fine particles are dominant, the coarse particles would occupy solid volumes originally occupied by the porous bulk volumes of the fine particles, thus increasing the packing density by means of *the occupying effect*.

If the size difference between the fine particles and the coarse particles is sufficiently large such that the presence of each type of particles would not affect the packing of the other type of particles, the binary mix of particles is said to have no interaction. However, if the size difference is not sufficiently large, the binary mix would have interactions between the particles. In this case, if the coarse particles are dominant, the packing of the coarse particles could be loosened by the addition of fine particles of which the size is not small enough to fit entirely into the voids of the coarse particles. This is known as *the loosening effect*. If, instead, the fine particles are dominant, then each and every coarse particle would be surrounded by a sea of fine particles which tend to have larger voids near the surfaces of the coarse particles. This is known as *the wall effect*. To account for the loosening and wall effects, two parameters, the loosening effect parameter a and the wall effect parameter b, are incorporated in many existing packing models.

With the filling, occupying, loosening and wall effects accounted for, the theoretical packing density of a binary mix of particles is derived as:

$$\phi_a = \frac{1}{\dfrac{r_1}{\phi_1} + \dfrac{r_2}{\phi_2} - (1-a)\dfrac{r_1}{\phi_1}} \tag{1}$$

$$\phi_b = \frac{1}{\dfrac{r_1}{\phi_1} + \dfrac{r_2}{\phi_2} - (1-b)\cdot\left(\dfrac{1}{\phi_2}-1\right)\cdot r_2} \tag{2}$$

in which ϕ_a denotes the packing density of the binary mix when the coarse particles are dominant; ϕ_b denotes the packing density of the binary mix when the fine particles are dominant; r_1 and r_2 are the volumetric fractions of the fine particles and coarse particles, respectively; and ϕ_1 and ϕ_2 are the packing densities of the fine and coarse particles, respectively. Equation (1) applies when the coarse particles are dominant whereas Equation (2) applies when the fine particles are dominant.

2.2 The Wedging Effect Parameter *c*

The theoretical predictions by Equation (1) and Equation (2) are plotted against the fine particles volumetric fraction r_1 in Figure 1. Generally, the curve representing Equation (1) and the curve representing Equation (2) intercept at the point of optimum fine particles volumetric fraction r_1^* which yields the maximum packing density. The interception of the two curves gives a sharp peak indicating high sensitivity of the packing density to the volumetric fractions when the volumetric fractions are close to optimum. However, in reality, the measured results, such as those plotted as data points in Figure 1, generally lie on a smooth curve which gives a lower sensitivity of the packing density to the volumetric fractions and, more importantly, a lower packing density than the theoretically predicted value when the volumetric fractions are close to optimum.

The discrepancy between the measured packing density and the predicted packing density is shown shaded in Figure 1, from which it can be seen that although the discrepancy is negligible when either the volume fraction of the fine particles is small or the volume fraction of the coarse particles is small, it is quite significant when the volume fractions of the fine and coarse particles are close to optimum. In general, the discrepancy is largest when the volume fractions are equal to the optimum volume fractions. As the optimum volume fractions occur when the amount of fine particles is just enough to fill the voids between the coarse particles, this reveals that the discrepancy is generally largest when the fine particles are just filling up the voids between the coarse particles.

The above phenomenon may be explained by *the wedging effect*. When coarse particles are dominant and the amount of fine particles is increasing, more and more fine particles would squeeze themselves into the voids between the coarse particles. However, when the voids between the coarse particles are close to being filled up, some of the fine particles would be trapped in the narrow gaps between the coarse particles, thereby wedging the coarse particles apart. As a result, the coarse particles are forced to attain a lower solid concentration in order to accommodate the fine particles, which somehow are not filling into the voids between the coarse particles but are acting like wedges at the gaps between the coarse particles. On the other hand, when fine particles are dominant and the amount of coarse particles is increasing, more and more coarse particles would enter into the sea of fine particles and the average thickness of the layers of fine particles surrounding the coarse particles would gradually decrease. As the thickness of the layer of fine particles surrounding a coarse particle becomes close to or even smaller than the size of the fine particles, the fine particles surrounding the coarse particle could become more like wedges inserted at the gaps between the coarse particles causing the solid concentrations of the fine particles near the surfaces of the coarse particles and the solid concentration of the coarse particles to be lower than predicted by any existing packing model.

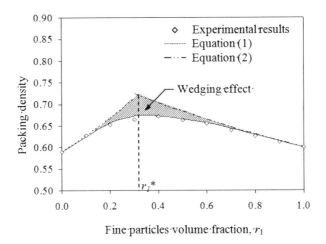

Figure 1 *Packing density against fine particles volume fraction*

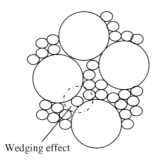

Figure 2 *The wedging effect*

Figure 2 illustrates the wedging effect when the fine particles are close to filling up the voids or slightly overfilling the voids between the coarse particles. The degree of such wedging effect can be represented by a wedging effect parameter c.

By equating Equations (1) and (2), the optimum fine particles volume fraction $r_1{}^*$ yielding the maximum packing density can be determined as:

$$r_1{}^* = \frac{(1-b) \cdot \left(\dfrac{1}{\phi_2} - 1\right)}{(1-a) \cdot \dfrac{1}{\phi_1} + (1-b) \cdot \left(\dfrac{1}{\phi_2} - 1\right)} \tag{3}$$

Assuming that the wedging effect is maximum when the volume fractions are optimum and that the wedging effect is proportional to $(r_1/r_1{}^*)^2$ or $(r_2/r_2{}^*)^2$, the packing density of the binary mix when the coarse particles are dominant may be derived as:

$$\phi_a = \frac{1}{\dfrac{r_1}{\phi_1} + \dfrac{r_2}{\phi_2} - (1-a) \cdot \dfrac{r_1}{\phi_1} \cdot \left[1 - c \cdot \left(\dfrac{r_1}{r_1{}^*}\right)^2\right]} \tag{4}$$

while the packing density of the binary mix when the fine particles are dominant may be derived as:

$$\phi_b = \frac{1}{\dfrac{r_1}{\phi_1} + \dfrac{r_2}{\phi_2} - (1-b) \cdot \dfrac{r_2}{\phi_2} \cdot (1-\phi_2) \cdot \left[1 - c \cdot \left(\dfrac{r_2}{r_2{}^*}\right)^2\right]} \tag{5}$$

Equations (4) and (5) together constitute a new 3-parameter particle packing model, in which the three parameters incorporated are the loosening effect parameter a, the wall effect parameter b, and the wedging effect parameter c. These three parameters are all empirical coefficients dependent on the size ratio r between the fine particles and the coarse particles. To determine their values, it is necessary to conduct packing density tests and then back calculate the respective values of the three parameters.

3   EXPERIMENTAL

An experimental program of testing the packing density of binary mixes of mono-sized particles at various volume fractions was launched to prove the theory of wedging effect and to determine the values of the three parameters a, b and c for the development of a new 3-parameter packing model. The particles used in the binary mix packing density tests were spherical glass beads of six mono-sized classes. By mixing two of the six size classes of glass beads, a total of fifteen series of binary mix were obtained, each of which with nine mix proportions. The steel cylindrical container used in the tests was the same as that stipulated in BS 812: Part 2:1995 for measuring the packing density of aggregate.[6] Dry packing with no compaction applied was adopted in the packing density tests of the binary mixes of glass beads. By plotting the measured packing density against the volume fraction and finding the best-fit curves based on Equations (4) and (5) using regression analysis, the values of the three parameters were determined.

4   RESULTS AND DISCUSSIONS

From the experimental results, it is found as expected that the wedging effect is negligible when the volume fraction of either the fine or coarse particles is small as there exists mainly the filling and loosening effect when the fine content is small and there exists vastly the occupying and wall effect when the coarse content is little. It is also found that the wedging effect increases when the volume fraction of fine particles approaches to r_1^*, which is the optimum volume fraction yielding the maximum packing density.

The empirical formulas for the three parameters a, b and c, each expressed in terms of the size ratio r, have been obtained by regression analysis of the back-calculated values of the three parameters, as presented in the following equations:

$$a = 1 - (1 - r)^{3.3} - 2.6 \cdot r \cdot (1 - r)^{3.6} \tag{6}$$

$$b = 1 - (1 - r)^{1.9} - 2 \cdot r \cdot (1 - r)^{6} \tag{7}$$

$$c = 0.322 \cdot \tanh(11.9 \cdot r) \tag{8}$$

The function of wedging effect at maximum packing density is exhibited in Figure 3. It is shown that after a rapid rise till the size ratio is around 0.1, the value of wedging effect parameter at maximum packing density eventually levels off to 0.322. This suggests that when the loosening effect is insignificant, the wedging effect at maximum packing density would be relatively less significant. Take an extreme case as an example, when the size ratio is nearly zero that ideally, there is no interaction between the particles, the maximum packing density would occur when the voids between the coarse particles are fully filled. In this case, there would be no wedging effect as the coarse particles are not pushed away by the fine particles at the maximum packing density. Therefore, as shown in Figure 3, when size ratio is zero, the wedging effect at maximum packing density would be zero. As the wedging effect begins only when the voids between the coarse particles is fully filled by the fine particles, the smaller interactions between the particles, the smaller wedging effect it would be at the maximum packing density provided that the interaction between the particles is very small, *i.e.* size ratio < 0.1.

On the other extreme that when the particles are totally interacting with each other, the filling effect and occupying effect are already offset by the loosening effect and wall effect respectively so that the maximum packing density of the binary mix is close to the single-sized packing densities of the constituent particles. In this case, the wedging effect is unimportant and can be set any arbitrarily value. It can be proved from Equations (4) or (5) that when there is total interaction between the particles, the values of loosening effect parameter a and wall effect parameter b would be equal to 1.0 such that the packing density is independent of the value of c, which for simplicity is just taken as 0.322.

In order to verify the accuracy of the 3-parameter particle packing model with wedging effect incorporated in predicting the packing density of a binary mix of particles, the published experimental results obtained from de Larrard[2] and Cintré[7] were extracted for validation. The prediction of packing density by the 3-parameter particle packing model is shown to be successful and accurate that the maximum absolute percentage error is between 1.36% and 1.43% with the mean absolute error between 0.55% and 0.66%. It is also found that the 3-parameter particle packing model can satisfactorily predict the packing density of binary mixes regardless of the compaction process.

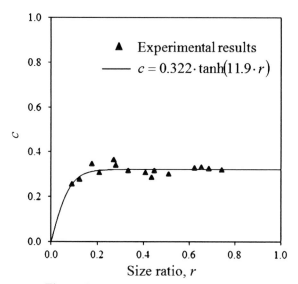

Figure 3 *Parameter c against size ratio*

5   CONCLUSION

A new kind of particle interaction, named as the wedging effect, has been identified. With the wedging effect incorporated into a new 3-parameter packing model, the accuracy of the packing model in predicting the packing density of a binary mixture can be improved. Furthermore, the new packing model with the wedging effect incorporated can provide a better insight into the structure of the packing of particles.

References

1   A. E. R. Westman and H. R. Hugill, *Journal of American Ceramic Society*, 1930, **13**, 767-779.
2   F. de Larrard, *Concrete Mixture Proportioning: A Scientific Approach*, E & FN Spon, New York, 1999, ch. 1, p.1.
3   R. F. Fedor and R. F. Landel, *Powder Technology*, 1979, **23**, 225.
4   R. K. McGeary, *Journal of American Ceramic Society*, 1961, **44**, 513.
5   K. Ridgway and K. J. Tarbuck, *Chemical Processing Engineering*, 1968, **49**, 103.
6   British Standards Institution, *BS 812 Testing Aggregates - Methods of Determination of Density*, London: BSI, 1995, part 2.
7   M. Cintré, *Rapport du LRPC de Blois*, 1988.

A DISCRETE ELEMENT MODEL FOR CONTACT ELECTRIFICATION

C. Pei,[1] D. England,[2] S. Byard,[3] H. Berchtold[2] and C.-Y. Wu[1]

[1] School of Chemical Engineering, University of Birmingham, Birmingham, B15 2TT, UK. E-mail: C.Y.Wu@bham.ac.uk
[2] Sanofi-Aventis Deutschland GmbH, Frankfurt, Germany
[3] Sanofi-Aventis, Northumberland, NE66 2JH, UK

1 INTRODUCTION

Contact electrification is referred to as the electrostatic charging process between objects during contacts. When two particles come into contact, the electrostatic charge can be transferred from one to another. Contact electrification[1-3] can occur during various powder processing operations, such as pneumatic conveying,[4-6] blending[7-9] and fluidization.[10-12] Once particles get charged through contact electrification, the attractive force induced by electrostatics can result in agglomeration and segregation, which will subsequently affect the quality of final products significantly. Therefore, understanding contact electrification processes is of fundamental importance to improve the process efficiency and quality of particulate products.

Charge transfer processes between particles have been experimentally investigated by many researchers.[9-15] For electrification during various powder handling processes, previous studies primarily concerned the relationships between specific process conditions and the charging phenomena. For example, LaMarche et al.[9] showed that the net charge of particles flowing from a cylinder increases linearly with the increase of the contact surface of the cylinder meanwhile the net charge density of the powder decreases from the outer annulus to the centre. As for gas-solid systems, such as fluidization and pneumatic conveying, the generation and distribution of charge highly depend on the flow regimes, relative gas velocity, particle size and geometry.[10,11] Although experimental studies on powder handling operations can reveal some mechanisms of charging processes, extrinsic conditions including humidity, contamination, temperature and even atmospheric pressure can significantly affect the electrification processes. Consequently, it is extremely difficult to achieve good reproducibility of experimental tests,[2,11,14,15] so that our understanding of contact electrification is far from complete.

In order to obtain a substantial understanding of contact electrification for theoretical analysis and modelling, contact charging processes during single and successive impacts between two objects have also been investigated.[16-22] For single impacts between an insulating particle and a surface, the charge of the particle highly relies on the impact velocity, maximum contact area and material properties. The charge will be accumulated on the surface of the particle and achieve an equilibrium state when the impact is repeated for the same process. When the particle slides along the surface with friction (i.e. sliding occurs), other factors, including the surface state and temperature during the sliding, can also affect charge transfer between materials.[23] Therefore, even for the collisions between two objects, the contact electrification process is a complicated event.

Based on the solid state theory, several theoretical models have been developed to explain the contact charging phenomena between metal-metal, metal-insulator and insulator-insulator materials.[24-27] In terms of physical process, it is believed that electron transfer dominates contact electrification between conductors whilst ion transfer or even material transfer is primarily responsible for the charging process between insulators. These mechanisms (electron transfer, ion transfer and material transfer) are likely involved simultaneously in reality, it is hence challenging to full understand the electrification processes. Therefore, semi-empirical concepts including work function and surface state models were proposed to analyze the charging processes.[1,2] Nevertheless, our understanding of the charging behaviour and dynamics of charged particles during industrial processes is still limited. Aiming to provide an insight into contact charging processes, a discrete element model is developed in this study, and the model is also validated using the experimental data reported in the literature.

2 THE DISCRETE ELEMENT MODEL

2.1 Discrete Element Methods

The Discrete Element Method (DEM) was initially proposed by Cundall and Strack[31] to solve dynamics of particle systems. In DEM, particles are treated as individual elements subjected to Newton's law of motion under the gravitational force and other interparticle forces. In order to calculate all the forces, the computational time is explicitly divided into small time steps, normally of micro-seconds, in which contact detection is performed to determine if a contact is established. Once a contact is established, contact forces between particles are calculated using a certain contact law and the kinematics of individual particle can then be determined. DEM can be used to calculate the trajectory and dynamics of each individual particle and provide specific information at the particle level in a powder handling process.[31]

DEM can also be modified to analyse electrification during collisions between particles. For instance, Watano et al.[5] used a simplified electrification model in DEM, in which the charge of a particle is proportional to the normal contact velocity and the number of collisions, to explore the contact electrification in powder pneumatic conveying process. But the saturation of charge on each particle during collisions was not considered. Hogue et al.[32] investigated the trajectories of particles rolling down an inclined plane using a time-dependent electrification model, which is more specific to triboelectrification rather than contact-dependent processes considered in this study.

In the present study, a condenser model based on the contact interaction in DEM and surface state of charged particles is established. The contact interaction, as the core of DEMs, is determined using classical contact mechanics.[33-35] For elastic particles, the normal contact is modelled using Hertz's theory while the theory of Mindlin and Deresiewicz is employed for the tangential interaction. The normal contact area is determined as follows:[35]

$$S = \pi a^2 = \frac{\pi a r_1 r_2}{r_1 + r_2} \tag{1}$$

where, S is the contact area, a is the contact radius, α is the relative approach (i.e. overlap) between particles; r_1 and r_2 are the radii of two particles in contact. Once the contact occurs, the charge, more precisely electrons, can be transferred from one surface to another because of the difference in electron affinity. According to the surface state model, for an insulating surface, the surface state can be expressed as:[1,18]

$$\emptyset = V + C_0 q \tag{2}$$

where, $\emptyset$ is the surface state of contact potential level; V is the work function; q is the net charge of the surface and the sign of charge q is also considered; C_0 is a constant related to the image effect. Thus, the contact potential difference (CPD) $\Delta\emptyset$ between two materials can be defined as:[1,18]

$$\Delta\emptyset = \emptyset_1 - \emptyset_2 \tag{3}$$

Based on the condenser model,[1,18] CPD is the driving force for electron transfer between contacting surfaces. For successive contacts between two different insulating materials, the charging process is governed by:

$$\frac{dq}{dn} = kS_{max}(\emptyset_1 - \emptyset_2) \tag{4}$$

where the transferred charge dq in contact n will move from $\emptyset_1$ to $\emptyset_2$ during the tunnel discharge, which is related to the constant k; S_{max} is the maximum contact area, which is calculated incrementally in DEM.

In the present DEM model, the maximum contact area for each pair of contacting particles is determined in each time step using Eq. (1). Once a contact is broken (i.e., the contacting particles separate), Eqs. (2)-(4) are used to calculate the current transferred charge. In reality, the contact point can be different from one collision to another,[17] which will lead to non-uniform charge distribution over the surface of the particle. As all particles are spherical in this study, it is assumed that the transferred charge distributes uniformly on the particle surface and will affect each contact charging process. The transferred charge on highly insulating particles can retain for several minutes to several hours.[36] The time period of retention is much longer than most powder handling processes involving intensive particle collisions. Therefore, the charge relaxation is ignored in this study.

2.2 Model set-up

Successive normal elastic impacts of a particle with a substrate were analysed using the DEM developed, in which the charge of the particle is determined using the condenser model described in Section 2.1. The model set-up is identical to the experimental one reported in Matsusaka *et al.*[18] as shown in Figure 1 for validation purpose. The insulating particle impacts the stationary and grounded conductive surface successively with various initial normal impact velocities v_0 ranging from 0.5 m/s to 4.0 m/s and initial net charges q_0 (-5 nC, 0 nC and 5 nC). The same material parameters for the particle and the substrate as those reported in Matsusaka *et al.*[18] are used and shown in Table 1. The transferred charge in each contact and the accumulation of the charge during the successive process are determined and analyzed.

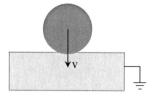

Figure 1 *The geometry set-up*

Table 1 *Material parameters*

	Particle	Substrate
Elastic modulus, E	2.0×10^6 Pa	210×10^9 Pa
Poisson's ratio, v	0.5	0.3
Density, ρ	890 kg/m^3	7800 kg/m^3
Diameter, D	31 mm	-
Material type	Rubber	Steel

3 RESULTS AND DISCUSSION

For the collision between an insulating particle and a grounded conductive substrate shown in Figure 1, the net charge on the substrate will dissipate instantaneously and remain zero. Hence, the transferred charge to the particle during each impact can be given as:

$$\frac{dq}{dn} = kS_{max}\left[V_s - \left(V_p + C_0 q\right)\right] \qquad (5)$$

where, V_s and V_p are the potentials related to the work functions of the surface and the particle, respectively. In the present DEM modelling, V_s is set to 4.52 V and V_p is 4.70 V, which are typical values of steel and polymer.[23] The parameter C_0 is related to the image effect and capacitance of the material and is set to 1.0×10^7 V·C^{-1}. The value of the parameter k is determined by fitting the experimental data of Matsusaka *et al.*[18] with Eq. (5) and $q_0 = 0$, and is 1×10^{-4} C·m^{-2}·V^{-1}, which is a typical value for the charge density during contact electrification between polymers and metals shown by many others.[18,20,21,37].

For successive collisions between the insulating particle and the grounded conductive substrate, the accumulation of charge can be derived from Eq. (5) as:

$$q = q_0 e^{-kC_0 S_{max} n} + q_\infty \left(1 - e^{-kC_0 S_{max} n}\right), \quad (n = 0, q = q_0) \qquad (6)$$

where

$$q_\infty = \frac{V_s - V_p}{C_0} \qquad (7)$$

q_0 is the initial charge and q_∞ is the equilibrium charge. For the given V_s, V_p and C_0, q_∞. is calculated using Eq.(7) and is equal to 18 nC.

Figure 2 shows the variation of transferred charge with maximum contact area for various initial charges q_0, in which the experimental results of Matsusaka *et al.*[18] are also superimposed. It is clear that the numerical results are in excellent agreements with the experimental ones. It is also clear that, for a given q_0, the transferred charge during each impact at various velocities is linearly proportional to the maximum contact area. It can also be seen that the transferred charge also depends on the initial charge. At the same impact velocity, the higher the initial charges, the less the charge is transferred. This is due to the fact that the potential difference varies with the net charge of the particle, according to Eq. (5).

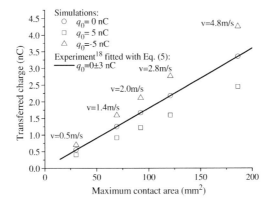

Figure 2 *The variation of transferred charge with maximum contact area*

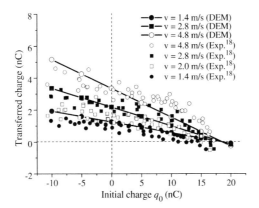

Figure 3 *The variation of transferred charge with initial charge at various initial impact velocities*

This is further illustrated in Figure 3, which shows the variation of transferred charge with initial charge at various initial impact velocities. The experimental data from Matsusaka *et al.*[18] are also superimposed. It can be seen that excellent agreements are obtained between numerical analysis and experimental study of Matsusaka *et al.*[18]. It is clear that the transferred charge decreases linearly with increasing initial charge, which is consistent with those shown in Figure 2. When the net charge on the particle is smaller than the equilibrium value q_∞, the particle will be positively charged. On the other hand, the particle will acquire negative charge once the net charge exceeds the equilibrium value q_∞.

Figure 4 presents the accumulation of electrostatic charge as a function of the number of impacts. It is clear that the charge of particle increases exponentially with the increase of the number of impacts. The charge of the particle eventually saturates to an equilibrium state after a few number of impacts. These results again agree very well with the experimental results of Matsusaka *et al.*[18] shown in solid symbols in Figure 4. For the experiments with larger impact intervals such as 20 or 30 seconds, the equilibrium value tends to be smaller. This is due to the fact that more charge relaxation to air takes place when the impact interval is longer. Therefore, it can be found that the charge at the equilibrium state depends on not only the initial CPD, but also the discharge and relaxation conditions, such as surface roughness and physical properties.

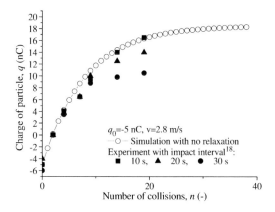

Figure 4 *The accumulation of electrostatic charge as a function of the number of impacts*

4 CONCLUSIONS

A discrete element model was developed for simulating successive contact electrification during powder processing. Validation of the developed model was also performed by comparing the numerical results with experimental data published in literature and excellent agreements were obtained, indicating that the developed DEM can be a feasible tool to explore dynamic charging behaviours of particles during powder processing. It was also demonstrated that, in each single collision, the transferred charge is proportional to the maximum contact area and decreases as the initial charge increases. In a successive process, the accumulation of charge on the particle is an exponential function of the number of collisions and eventually reaches an equilibrium state.

Acknowledgement

This project is fully funded by Sanofi-Aventis.

References
1   S. Matsusaka, H. Maruyama, T. Matsuyama, M. Ghadiri, *Chem. Eng. Sci.*, 2010, **65**, 5781.

2   L.B. Loeb, *Science*, 1945, **102**, 573.
3   J. Lowell, A.C. Roseinnes, *Adv. Phys.*, 1980, **29**, 947.
4   J. Yao, Y. Zhang, C.H. Wang, Y.C. Liang, *AIChE J.*, 2006, **52**, 3775.
5   S. Watano, S. Saito, T. Suzuki, *Powder Technol.*, 2003, **135**, 112.
6   J. Yao, Y. Zhang, C.H. Wang, S. Matsusaka, H. Masuda, *Ind. Eng. Chem. Res.*, 2004, **43**, 7181.
7   K.C. Pingali, T. Shinbrot, S.V. Hammond, F.J. Muzzio, *Powder Technol.*, 2009, **192**, 157.
8   D.A. Engers, M.N. Fricke, R.P. Storey, A.W. Newman, K.R. Morris, *J. Electrost.*, 2006, **64**,826.
9   K.R. Lamarche, X. Liu, S.K. Shah, T. Shinbrot, B.J. Glasser, *Powder Technol.*, 2009, **195**, 158.
10 X.B. Yu, W. Li, Y. Xu, J.D. Wang, Y.R. Yang, N. Xu, et al., *Ind. Eng. Chem. Res.*,2010, **49**, 132.
11 J. Guardiola, V. Rojo, G. Ramos, *J. Electrost.*, 1996, **37**, 1.
12 S.L. Escalante, G. Touchard, G. Dominguez, *2002 Annual Report Conference on Electrical Insulation and Dielectric Phenomena*. 2002, 694.
13 A.G. Bailey, *Powder Technol.*, 1984, **37**, 71.
14 M.D. Hogue, C.R. Buhler, C.I. Calle, T. Matsuyama, W. Luo, E.E. Groop, *J. Electrost.*, 2004, **61**, 259.
15 S. Kittaka, *J. Phys. Soc. Jpn.*, 1959, **14**, 532.
16 H. Masuda, K. Iinoya, *AIChE J.*, 1978, **24**, 950.
17 T. Matsuyama, H. Yamamoto, *IEEE Trans. Ind. Appl.*, 1995, **31**, 1441.
18 S. Matsusaka, M. Ghadiri, H. Masuda, *J. Phys. D: Appl. Phys.*, 2000, **33**, 2311.
19 T. Matsuyama, M. Ogu, H. Yamamoto, J.C.M. Marijnissen, B. Scarlett, *Powder Technol.*, 2003, **135-136**, 14.
20 T. Matsuyama, H. Yamamoto, *Chem. Eng. Sci.*, 2006, **61**, 2230.
21 H. Watanabe, M. Ghadiri, T. Matsuyama, Y.L. Ding, K.G. Pitt, H. Maruyama, et al., *Int. J. Pharm.*, 2007, **334**, 149.
22 G. Rowley, *Int. J. Pharm.*, 2001, **227**, 47.
23 J.A. Cross, *Electrostatics: Principles, Problems and Applications,* Adam Hilger, Bristol, 1987, ch. 2, p. 17-45.
24 L.B. Schein, M. Laha, D. Novotny, *Phys. Lett. A*, 1992, **167**, 79.
25 G.S.P. Castle, L.B. Schein, *J. Electrost.*, 1995, **36**, 165.
26 C. Y. Liu, A.J. Bard, *Chem. Phys. Lett.*, 2009, **480**, 145.
27 S. Matsusaka, H. Masuda, *Adv. Powder Technol.*, 2003, **14**, 143.
28 M. Hennecke, R. Hofmann, J. Fuhrmann, *J. Electrost.*, 1979, **6**, 15.
29 H. Masuda, T. Komatsu, K. Iinoya, *AIChE J.*,. 1976, **22**, 558.
30 N. Duff, D.J. Lacks, *J. Electrost.*, 2008, **66**, 51.
31 P.A. Cundall, O.D.L. Strack, *Geotechnique*, 1979, **29**, 47.
32 M.D. Hogue, C.I. Calle, P.S. Weitzman, D.R. Curry, *J. Electrost.*, 2008, **66**, 32.
33 C.Y. Wu, *Finite element analysis of particle impact problems*, The University of Aston, 2001.
34 Y. Guo, K.D. Kafui, C.Y. Wu, C. Thornton, J.P.K. Seville, *AIChE J.*, 2009, **55**, 49.
35 K.L. Johnson, *Contact mechanics*, Cambridge University Press, Cambridge, 1985.
36 M.I. Kornfeld, *J. Phys. D: Appl. Phys.*, 1976, **9**, 1183.
37 N. Masui, Y. Murata, *Jpn. J. Appl. Phys., Part 1*, 1983, **22**, 1057.

LARGE-SCALE DISCRETE ELEMENT SIMULATIONS OF GRANULAR MATERIALS

J.G. Liu[1], Q.C. Sun[1], Q.K. Liu[2] and F.Jin[1]

[1] State Key Laboratory of Hydroscience and Engineering, Tsinghua University, China
[2] Institute of Applied Physics and Computational Mathematics, Beijing, China

1 INTRODUCTION

The mechanical behaviour of granular materials is important in both the engineering and physics communities. Discrete element method (DEM) simulations could provide detailed information about the motion and the contact force of individual particles as they interact with each other and the boundary.[1, 2] The macroscopic properties are then determined by appropriate space and time averaging methods. However, DEM simulation is computationally intensive, which limits either the length of a simulation or the number of particles to be used. Most simulations have to be confined to two dimensions. Therefore, a large-scale parallel computation framework should be developed.[3]

We have initiated the development of parallel DEM programs base on the JASMIN infrastructure, which take advantage of parallel processing capabilities to scale up the number of particles or the length of simulation. The parallel implementation of DEM is briefly introduced firstly, and the preliminary results of an isotropically compression simulation are reported. The relationship between wall pressure and packing fraction is examined.

2 THE JASMIN FRAMEWORK FOR GRANULAR MATERIALS

The JASMIN (J parallel Adaptive Structured Mesh applications INfrastructure) is developed by Beijing Institute of Applied Physics and Computational Mathematics. JASMIN supports innovation of computing methods and high performance algorithms, and assists the developing of parallel code scaling to thousands of processors. JASMIN has a user-friendly interface, and hides the implement details of high performance parallel computing and adaptive computing from users.

The complexity of application systems and that of computer architectures are major challenges for parallel programming in the field of scientific computing. Programming paradigms are waiting for innovations for multidisciplinary intersection and cooperation. JASMIN succeeds in the field of scientific computing for the adaptive structured meshes, and supports many important applications including fluid dynamics, radiation transportations, particle simulations and so on.

Figure 1 depicts a typical data structure in JASMIN. Figure 1a shows a two-dimensional structured mesh consisting of 20×20 cells. It is decomposed into seven patches and each patch is defined on a logical index region named by Box. In Figure 1b, these patches are ordered and are distributed between two processors. The left four patches belong to one processor and the right three patches belong to another processor. In Figure 1c, the sixth patch is shown to illustrate the neighbour relationships. It is the neighbourhood of the other four patches and contacts with the physical boundaries. [4]

To store the transferred data from neighbouring patches, each patch extends its box to a ghost box within a specific width. Patch-based data structures are suitable for the cache based memory hierarchy. For a given cache memory size, we can adjust the size of patches to improve the cache hit ratio.[5]

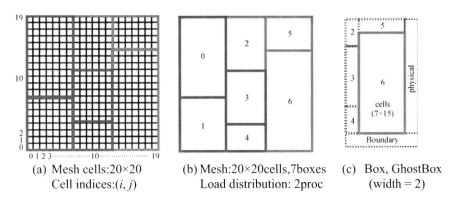

(a) Mesh cells:20×20
Cell indices:(i, j)

(b) Mesh:20×20cells,7boxes
Load distribution: 2proc

(c) Box, GhostBox
(width = 2)

Figure 1 *A typical data structure in JASMIN*

Figure 2 shows the particles assigned in process, patch and cells. In the discrete element simulations, we put particles into mesh cells. There is one particle at most in each cell. We can use the cell's index to identify particles, such as location, velocity, mass and materials properties. A patch is then generated as a rectangular block including a group of particles. A corresponding parallel algorithm is automatically implemented with the JASMIN framework. The algorithm deals with dynamical load balancing, particles migration and data exchange among these objects. In our simulations, a three-dimensional packing is divided into 8 computing processors, 700 data patches, and 18000 cells, as shown in Figures 2a, 2b and 2c, respectively. Figures 2d, 2e and 2f show the related particles in one processor, one data patch and one cell, respectively.

3 RESULTS AND DISCUSSION

The first work with this improved framework is the simulation of a granular assembly under isotropic compression. The Hertzian interaction potential between particles is used. The Young's modulus is $E = 6 \times 10^8$ Pa, surface friction μ=0.5, and density $\rho = 2600\text{kg/m}^3$. The system contains 1.02×10^6 particles with uniformly distributed radii between 0.35 and 0.65 mm. The particles are randomly distributed in a cylinder. Its diameter is 10 cm, and the height is 20 cm, as schematically shown in Figure 3.

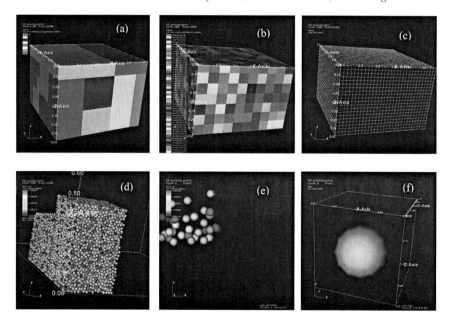

Figure 2 *Particles allocation in processes, data patches and cells. The upper row is the
data structure and the lower row shows particles in a process, a data patch and
a cell.*

There are two stages during the simulation. Firstly, we reduce the radius of particles
and then enlarge them in order to eliminate crystallization. The initial packing fraction is
0.3447. Secondly, we keep proportionally enlarging the particle size until the packing
fraction reaching 0.7. The growing speed is 0.0001 times of particle's original size per 10
steps. The time step size is 6.8×10^{-6} s, and the total step is 1.4×10^{5}. In the simulation, we
calculate 1.02×10^{6} particles with 512 processors, and it costs 0.024 s per time step. The
results can be seen in Figure 4.

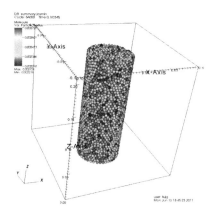

Figure 3 *A snapshot of particle packing.*

Wall pressure P is one of the key parameters to measure the structure of granular packing. A system is mechanically stable only when P is not zero. Figure 4 shows the variation of boundary pressure P with the packing fraction ϕ. As $\phi < 0.58$, P is always close to zero; when $\phi > 0.58$, P sharply increasing., so the critical packing fraction is $\phi_0 = 0.58$. But along with the increase of ϕ, the value of P increases with many sudden drops, i.e. rapid stress relaxations, which indicates particles re-arrangements, as shown in the inset of Figure 4. The scaling law is $P \sim (\phi - \phi_0)^{0.829}$ at the first stage, and $P \sim (\phi - \phi_0)^{1.762}$ at the second stage. That is because in the first stage the contact area is small, so the normal force follows Hertz model. But when the packing fraction is large, as in the second stage, the contact area is no longer small, and the friction effect is enhanced. The above findings suggest that the scaling law of frictional soft sphere systems is determined by the distance from the point of the system becoming stable. Accurate relationships and reasonable explanation of the scaling law need further investigation.

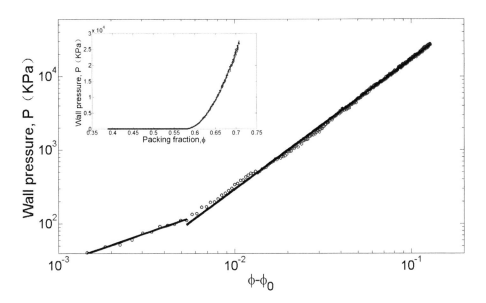

Figure 4 *Boundary pressure P vs. ($\phi - \phi_0$). The inset shows the variation of P with ϕ. Open circle donates simulation value, and black solid line is the fit.*

4 OUTLOOK

Our aim is to establish a large scale parallel computation platform for granular materials based on the super computers and JASMIN framework. We intend to analyse their internal structures and evolutions, and find the relationship between macroscopic mechanical behaviour and microscopic physical structures. The next step would focus on the controlling the stress state in the sample during the test. The servo method would be employed.

References

1 H.P. Zhu, Z.Y. Zhou, R.Y. Yang and A.B. Yu, *Chemical Engineering Science*, 2008, **23**, 63.
2 J.H. Walther and I.F. Sbalzarini, *Engineering Computations*, 2009, **6**, 26.
3 I.F. Sbalzarini, J.H. Walther, M. Bergdorf, S.E. Hieber, E.M. Kotsalis and P. Koumoutsakos, *Journal of Computational Physics*, 2006, **2**, 215.
4 Z.Y. Mo, A.Q. Zhang, X.L. Cao, Q.K. Liu, X.W. Xu, H.B. An, W.B. Pei and S.P. Zhou, *Frontiers of Computer Science in China*, 2010, **4**, 4.
5 X.L. Cao, Z.Y. Mo, X. Liu, X.W. Xu and A.Q. Zhang, *Science in China*, 2011, **4**, 54.

DISCRETE ELEMENT MODELLING OF A MOLE SAMPLING MECHANISM

S. Zigan and A. Hennessey

Chemical & Process Engineering, University of Surrey, Guildford, GU2 7XH, UK

1 INTRODUCTION

The research on the composition of soils in outer space e.g. on the planet Mars requires the sampling and analysis of materials. For sampling soils in certain depths a compact mechanism has to penetrate the soil by compressing and displacing materials. Soil penetration mechanisms such as the mobile 'penetrometer' have an integrated 'Mole' sampling mechanism (MSM).[1] The design of a MSM is described in detail by Richter et al.[1]

Experimental testing of the MSM under Earth environment is a complex task because parameters such as the gravity can hardly be adjusted to Mars environment. Another way of testing the MSM is to develop a simulation model which is validated with experimental data from Earth. Again, parameters have to be adjusted according to outer space, e.g. Mars environment.

The development of a simulation model incorporating the soil behaviour e.g. the particle-particle interaction is based on the discrete element method (DEM).[2] In DEM each particle is modelled as a discrete object and associated with a geometry and a surface topology. Usually, DEM models require large computational resources which result in long simulation times because the particle movements and interactions are tracked over time.[3]

The aim to have a fast and simple DEM model requires either significantly increasing the particle size of the soil or using a small representation of the geometry. The latter was chosen because realistic soil characteristics such as shear resistance, compression and elasticity of the material could be incorporated.[2]

The aim of the research is to develop a DEM model which predicts the compressive forces on the MSM at certain bed depths. One objective is to keep most of the material parameters in the simulation model constant, so that results such as the compression force become comparable for different soil depths.

2 MODEL DESCRIPTION

The set up of the DEM model requires the input of the geometry data and particle parameters. The basic geometry is a box to hold the soil particles. The size of the model is reduced to a 'slice' of the width 5 cm x 0.5 cm, and depth 1 cm. The width of the model required is determined by the soil properties and configured to take into account the soil compression. The volume of the box which is filled with particles is approximately $1.5cm^3$, when the volume of the MSM shaft is subtracted. Quartz sand and flintstone (QS-FL8) are used as reference material and parameters such as the Poisson's ratio result from experiments. According to Das[4] the Poisson's ratio is determined by triaxial testing, and is defined as the ratio of change in lateral strain to axial strain. Additional input parameter for the DEM model were the Rayleigh time step 1.23E-06 s and the fixed time step which was 70% of the Rayleigh time step. The cell size was 3 times R_{min} (375 μm) and the number of cells was 142,912.

3   METHODOLOGY

The MSM is implemented in the model as a 3D geometry and imposes a mechanical impact on the soil. The impact on the soil depends on the movement of the MSM. As a result a force will be created which can be measured and plotted for various depths. The results obtained from the simulation model are compared with experimental data in a test bed at a depth of around 30cm.

The MSM as shown in Figure 1, has an entire length of 270 mm, and a total width of 20 mm. The 3D geometry for the simulation model has a conical shape for the tip of 10 mm radius, and 20 mm length, with the shaft at 10 mm radius, and length of up to 250 mm.

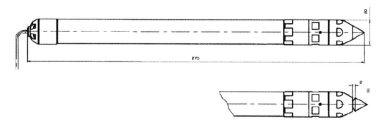

Figure 1   *Top: MSM sampling device closed; bottom: MSM sampling device open, revealing annular gap of 2mm width to allow lateral soil ingress (Source: 3D geometry was supplied by the German aerospace centre DLR).*

Table 1   Material properties for the DEM simulation

	Particle – soil (sand QS)	MSM steel
Particle size distribution (5% - 95%)	250μm - 500μm	
Modal particle size	420μm^2	
Poisson's ratio	0.13	0.3
Shear modulus	5.5E+07 Pa	7.93E+10 Pa
Density	1.45 g/cm^3	8.03 g/cm^3
Coefficient of restitution	0.5	0.95
Coefficient of static friction	0.6	0.5
Coefficient of rolling friction	0.1	0.01

The particles are quartz sand and Flintstone (QS-FL8), with a range of fine grains, the significant size range being 10µm to 500µm. Parameters from the quartz sand and mole are presented in Table 1. The particle size distribution of the quartz sand and Flintstone mixture used as reference material for the simulation is shown in Figure 2.

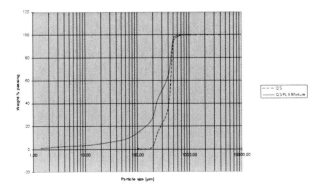

Figure 2 *Sieving results for MSM test soil ('QS-FL8 Mixture'), in comparison with dry quartz sand ('QS').*

The bulk density of the quartz sand and Flintstone mixture at a depth of 30 cm was calculated using Equation 1 and resulted in a value of 1.45 g/cm^3.[1]

$$\rho = x * \frac{z+12.2}{z+18} \tag{1}$$

where

ρ = density of the soil at specified depth (g/cm3)
x = maximum soil density (g/cm3)
z = depth in the soil (cm)

By including the density values from equation 1 in the simulation model a density profile was obtained which matched the soil density in the test bed at 30 cm. After completing the simulation, the compressive force was analysed and compared with experimental data.[1] In order to get a singular result, the simulation data were grouped to match the tip and shaft area of the MSM and adjusted by choosing values with maximum exposure to particles. After 1 sec simulation time the run was stopped because the mole penetrated the box completely.

4 RESULTS AND DISCUSSIONS

Results from the simulation showed that the forces in the soil bed and MSM can be calculated after post processing the data from the simulation. The force calculation had to take into account the contact area of the MSM tip and shaft with the particles in the 'slice' geometry. This post processing is necessary because the MSM tip and shaft are only partially in contact with the particles. This approach reduced the variability of the results because only data showing a maximum contact area of the MSM with the soil were considered. Results of the simulation are shown in Figure 3.

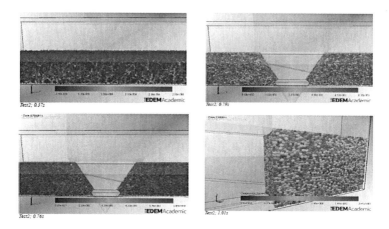

Figure 3 *Simulation result after different simulation times*

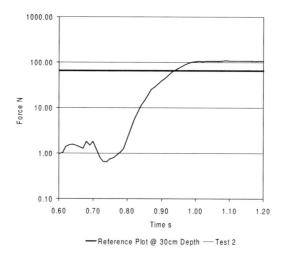

Figure 4 *Plot of the compressive force of the tip and shaft over simulation time*

After post processing simulation data compressive forces on the MSM were obtained which were compared to experimental data, see Figure 4. The compressive force data were collected on the MSM tip and shaft. The values for the compressive force predicted by the simulation are slightly larger than the experimental data presented by Richter et al.[1] (reference plot). It can be noticed that the compressive force became constant in the simulation after the shaft of the MSM completely penetrated the particle bed of the 'slice' geometry. The used density definition for the soil in the simulation matched the pressure profile in the experimental test bed at 30 cm. Thus, the bulk characteristic of the quartz sand could be well reproduced in the simulation.

5 CONCLUSIONS

The 'slice' geometry approach showed a good agreement with experimental data obtained in a test bed at 30 cm. The advantage of this simulation model was that real particle properties and bulk characteristics were used. Another advantage was the fast convergence of the simulation and the reduction of the model to a one variable problem. Fixing the particle size in the simulation model ensured that the Rayleigh time step remained unchanged.

Future work will be concerned with investigating how the accurate particle mass can be calculated for the geometric box to obtain the desired bulk density. This is a necessary step to understand how DEM can be used to simulate the MSM drilling process at different bed depths.

Acknowledgements

Many thanks to the University of Surrey for the facilities, students and staff alike. Especial thanks to the German Aerospace Centre (DLR) who supported this project.

References

1   L. Richter, V. Gromov, & E. Re, *Mole Sampling Mechanism (MSM), Final Report of ESA contract 14233/00,* 2006, 39.
2   N. Belheine, J.-P. Plassiard, F.-V. Donze, F. Darve and A. Seridi, *Comp. and Geotech.,* 2009, **36**, 320.
3   R.-M. O'Connorl, J.-R. Torczynski, D.-S. Preece, J.-T. Klosek, J-R. Williams, *Int. J. Rock Mech. & Min. Sci.,* 1997, **34**, 231.
4   B.M. Das, *Advanced Soil Mechanics,* 2nd edn., London: Taylor & Francis, 2002, pp. 343-344.

STUDY OF GRANULAR MATERIAL MULTI-SCALE CHARACTERISTICS FROM COMPLEX NETWORK PERSPECTIVE

J. Wu, F. Liu, Y. Dai, Z. Wang and D. Hu

School of Chemical Machine Engineering, Dalian University of Technology, Dalian, Liaoning, 116024, China. Email: wujt75@dlut.edu.cn

1 INTRODUCTION

Granular materials are widely used in the chemical industry, in particular, reaction engineering, production of powders, storage of grains, particle separation and granulation. Approximately 1/2 of the products and about 3/4 of the raw materials of chemical industry are in the form of granules.[1] Granular materials are widely studied since they exhibit unusual and distinct properties. Considerable importance has been given in industries for handling of granular materials. But thorough understanding of the flow behaviors of granular materials is not well established. Previous work on particle flow in silos is mainly at the macroscopic or bulk scale,[2-4] which is helpful in developing a broad understanding of granular flow in silos. But effects of individual particle properties on granular bulk flow are significant but not well understood. To overcome this problem, both experimental investigation and numerical simulations have been performed to explore the dynamics of granular flow at a microscopic or particle scale.

Force networks in the granular materials form the skeleton of static granular matter. They are the key factor that determines mechanical properties such as stability, elasticity and sound transmission, which are crucial for civil engineering and industrial processing.[5] The force networks can be analyzed based on complex network. In this paper, a multi-scale structural study of granular flow based on complex network is performed to understand the fundamental mechanism.

2 DEM MODELS

The principle of DEM is to track the movement of each particle, in a time stepping simulation, the trajectory and rotation of each element in a system are calculated to evaluate its position and orientation, and then to detect the interactions between the elements themselves and also between the elements and their environment. The interaction will then subsequently determine the new position of each element. Conventional DEM has no additional limits on the rolling of particles. Oda and Iwashita[6, 7] considered the effect of rolling resistance at contact point, and developed a modified DEM (i.e., MDEM). But it is still arguable to use rolling friction models in DEM.[8] In silo, the flow behaviour of particles is slow-flow. The rotation of particles is hence negligible. So the rolling friction is not taken into account in this study.

In Table 1, the simulation parameters are listed. Samples with four types of size distributions (random distribution, normal distribution, binary mixing, and uniform diameter) with the same average diameter of 0.8 mm are considered in this study, which are shown in Figure 1. Four thousand particles are generated and gradually settled onto the silo to form a packing under gravity, and are then discharged when the outlet is opened.

Table 1   *The properties of granular materials*

Parameter	Value
Number of particles, N	4000
Particle diameter, d_p (mm)	0.6-1.0
Particle density, ρ_p (kg·m^{-3})	946
Friction coefficient	
Particle/particle, μ_p	0.52
Particle/wall, μ_w	0.43
Spring coefficient	
Particle/particle, S_p (N·m^{-1})	5289
Particle/wall, S_w (N·m^{-1})	10580
Damping coefficient	1
Particle/particle, c_p [N·(ms^{-1})$^{-1}$]	1.39
Particle/wall, c_w [N·(ms^{-1})$^{-1}$]	2.78
Time step, Δt (s)	2×10^{-6}

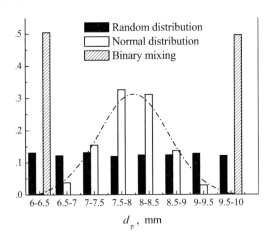

Figure 1   *The particle size distributions (PSD) considered*

The DEM involves three key parameters: spring stiffness (k), damping coefficient (c), and friction coefficient (μ). These parameters can be obtained from experiments.[9] In this study, the parameters obtained by Yang and Hsiau[1] are used.

3 COMPLEX NETWORK FOR GRANULAR MATERIALS

A network or graph is a collection of nodes and links. At any given strain state, a (complex) contact network can be constructed in which the particles are cast as the nodes of the network and a connecting link or edge exists between two particles that are in contact. Particle rearrangements in the quasi-statically deforming granular material will be reflected in the evolving network: some old network connections break as contacts are lost, while some new connections form as contacts are created. The contact network is an unweighted, undirected graph and is usefully summarized by an adjacency matrix. Complex network offers a multitude of statistical measures for the quantitative characterization of evolving networks, many of which can be calculated by manipulating the adjacency matrix. In two-dimension a granular contact network is a planar graph whose properties can be readily explored. Nevertheless, more useful information can be obtained from the complex network for three-dimensional granular systems.[10]

Network theory is the study of mathematical structures used to model complex relations between subjects from a certain collection. In this work, the contact topology of granular materials can be treated as a complex network where particles are *nodes* and the interacting force (contact) pairs are *edges*. The degree distribution, network diameter, clustering coefficient and so on are used to analyse the multi-scale structures in granular materials.[11]

Figure 2 shows the particle contact networks for different PSDs. The networks for mono-disperse systems are rather regular. The networks are disordered for the poly-disperse system. For mono-disperse system the particles packed as a hexagonal or quadrangular mode. So there are many triangles and quadrilateral with same sizes in the network. For poly-disperse systems, the structure of the packed particles was irregular. There are some larger holes in the network for the random distribution system and smaller triangles in the network for the binary mixture system.

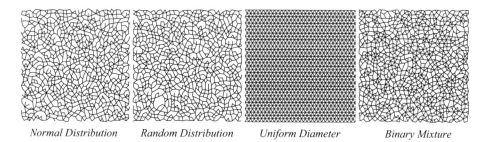

| *Normal Distribution* | *Random Distribution* | *Uniform Diameter* | *Binary Mixture* |

Figure 2   *The contact network for different PSDs*

4 RESULTS AND DISCUSSION

4.1 Macroscopic Flow Patterns

Two length scales, the macroscopic and microscopic levels, are examined. At the macroscopic level, the flow patterns of particles are analysed from DEM simulations and physical experiments. A model flat-bottomed silo with no insert is used to validate the DEM simulations and to examine the flow pattern by comparing experimental results with simulations.

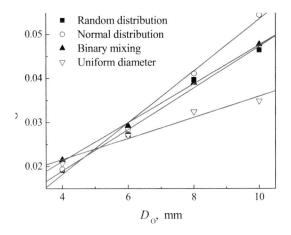

Figure 3   *The relationship between the discharge rate and the orifice size for different PSDs*

The variation in particle velocities may lead to the variation of discharge rate. Therefore, the geometric and physical parameters also affect the discharge rate of hoppers. Beverloo *et al.* [2] in 1961 plotted $G^{2/5}$ against D_o and obtained a linear relationship, and correlated their results in the form $G^{2/5}= C(D_o-Kd_p)$ for mono-disperse systems, where G is the discharge rate, C is a constant related to the physical properties of the flow such as bulk density, and K is a coefficient related to the particle shape. The relationship has been found accurate by many subsequent workers. Our results were in good agreement with this correlation, as shown in Figure 3, in which the discharge rate increased with the orifice size both for the mono-disperse system and for the poly-disperse system. The orifice size had a limited impact on the mono-disperse system, while it had a significant impact on the normal-disperse system in flat-bottomed silo. When the orifice was smaller, the larger particles in the poly-disperse system likely resulted in "bridging action" in the vicinity of the orifice, which led to a lower discharge rate for the poly-disperse system compared to the mono-disperse system. With the increase in orifice size, the "bridging action" became weaker, and the mixing of poly-disperse particles propitious to granular flow, which resulted in a higher discharge rate for the poly-disperse system than that for the mono-disperse system.

4.2 Microscopic Analysis Based on Complex Networks

At the microscopic level, the spatial and statistical distributions of micro-dynamic variables related to PSDs were established. A network efficiency E can be defined as

$$E = \frac{1}{N(N-1)} \sum_{i \neq j} \frac{1}{d_{ij}} \tag{1}$$

where the sum takes all pairs of vertices into account. This measurement quantifies the efficiency of the network in conveying information between vertices, assuming that the efficiency for conveying information between two vertices i and j is proportional to the reciprocal of their distance. The networks efficiency for different PSDs is listed in Table 2. The mono-disperse granular system had maximum networks efficiency, indicating more restricted in particle flow. It can be explained that the discharge rate for mono-disperse system is increases slowly with the orifice size.

Table 2   *The networks efficiency for different PSD*

PSD	Binary mixture	Normal distribution	Random distribution	Uniform distribution
E	5.373	4.988	5.281	5.409
$<k>$	5.3	3.9	4.7	5.8

The degree of a vertex i, hence k_i, is the number of edges connected to that vertex. For undirected networks it can be computed as

$$k_i = \sum_j a_{ij} = \sum_j a_{ji} \tag{2}$$

The degree is an important characteristic of a vertex.[12] Based on the degree of the vertices, it is possible to derive many measurements for the network. Additional information is provided by the degree distribution, $P(k)$, which expresses the fraction of vertices in a network with degree k. In our case, the degree k of a node represents the number of contacts between neighbouring particles. Then, the degree distribution $P(k)$ is the distribution function of the number of contacts per particle.

Figure 4a shows the degree distribution for difference PSDs. For the mono-disperse system the degree is a δ distribution with a peak at 6. This implies that the particle contact network of the mono-disperse system is a regular network. For the poly-disperse system, the degree is a quasi-Poisson distribution, in which the degree has a peak at the average k ($\langle k \rangle$ as shown in Table 2) for binary mixture and random distribution. The contact force networks for poly-disperse systems are essentially the small world networks featured with a Poisson degree distribution and a small average degree $\langle k \rangle$.

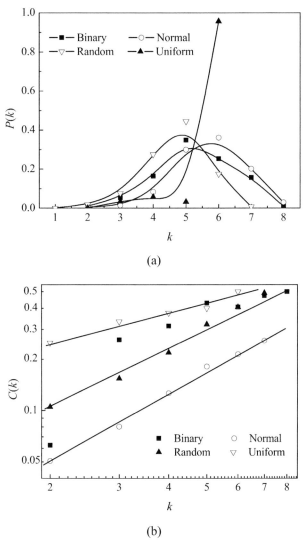

(a)

(b)

Figure 4 The effects of PSD on complex networks of granular materials (a) The degree distribution; (b) The relationship between the degree and clustering coefficients

The third fundamental quantity of a network which is often considered is the clustering coefficient[13]. The clustering coefficient is a measure of the number of triangle motifs, closed paths of length three, i.e., 3-cycles, in a network and it ranges from 0 (no link between a nodes neighbours) to 1 (nodes and their neighbours form a clique). A high value of clustering coefficient for an individual particle corresponds to a local closely packed structure whereas a low value means fewer 3-cycles, locally loosely packed and less connectivity. An arrangement of three particles in the form of a triangle is known to result in frustrated rotations and hence stability. Since the clustering coefficient is a measure of the amount of triangles in the contact network, a reduction in its value corresponds to a loss of these stabilizing structures in the material.

In the graph theory, a clustering coefficient is a measure of degree to which nodes in a graph tend to cluster together. It can be defined as:

$$C_i = \frac{2l_i}{k_i(k_i - 1)} \tag{3}$$

Figure 4b shows the relationship between the degree and clustering coefficients. The degree and clustering coefficients ($C(k)$ is the average clustering coefficient for all the particles with the degree of k) of network are in a linear relationship in log-log coordinates for normal, random, and uniform distributions. It indicates that there were hierarchical structures in networks.[14]

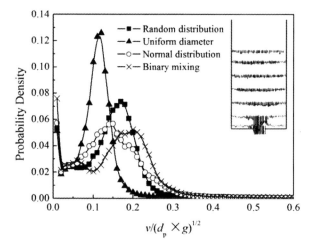

Figure 5 *Probability distribution of particle velocity for different PSDs*

Figure 5 illustrates the probability density distribution of particle velocity for different PSDs. The probability density distribution curves of the vertical velocity have two peaks. The first peak at a very small value corresponds to the particles in the "dead zones", where particles are packed closely. The second peak at a larger value is related to the particles of "mass flow" in the central region, where particles packed loosely. There are different structures in the system, and the difference in structures has great effects on particle flow.

5 CONCLUSIONS

DEM simulations and a complex network method were used to investigate the multi-scale characteristics of granular materials. At the macroscopic level, the flow patterns of particles were investigated numerically and experimentally. The relationship between the discharge rate and the orifice size can be described by the Beverloo equation, i.e. $G^{2/5}$ against D_o as a linear relationship for the mon-disperse system. And the particle size distribution has great effects on granular flow. The mixing of poly-disperse particles was propitious to granular flow.

At the microscopic level, the characteristics of granular materials were explored using complex networks. The spatial and statistical distributions of micro-dynamic variables related to PSD were established based on network efficiency, degree distribution, and clustering coefficient. The particle contact network of the mono-disperse system is a regular network, while the contact networks for the poly-disperse system are the small world networks. There are hierarchical structures in particle contact networks, and the difference in structures has great effects on particle flow. The macroscopic flow patterns are related to the structural characteristics of granular networks.

Nomenclature

a	number of edges	G:	discharge rate
c:	damping coefficient	l:	edges that actually exist between the k_i nodes
C:	clustering coefficients	k:	degree
d_p:	particle diameter	Δt:	time step
D_o:	diameter of offset	ρ_p:	particle density
E:	network efficiency	μ:	friction coefficient

References
1. S.C. Yang and S.S. Hsiau, *Powder Technol.*, 2001, **120**, 244.
2. R.M. Nedderman and U. Tuzun. *Chem. Eng. Sci.*, 1982, **37**, 1597.
3. U. Tuzun, G.T. Houlsby, R.M. Nedderman and S.B. Savage. *Chem. Eng. Sci.*, 1982, **37**, 1691.
4. S.B. Savage, R.M. Nedderman, U. Tuzun and G.T. Houlsby, *Chem. Eng. Sci.*, 1983, **38**, 189.
5. S. Ostojic, E. Somfai and B. Nienhuis, *Nature*, 2006, **439**, 828.
6. M. Oda and K. Iwashita, *Int. J. Eng. Sci.*, 2000, **38**, 1713.
7. K. Iwashita and M. Oda, *Powder Technol.*, 2000, **109**, 192.
8. H.P. Zhu and A.B. Yu, *Powder Technol.*, 2006, **161**, 122.
9. B.P.B. Hoomans, *Granular dynamics of gas–solid two-phase flows*, Twente University, Enschede, 2000.
10. A. Tordesillas, P. O'Sullivan, D.M. Walker and C. R. Mecanique, 2010, **338**, 556.
11. L. da F. Costa, F. A. Rodrgues, G. Traveso and P.R.V. Boas, *Adv. Phys.*, 2007, **56**, 167.
12. L.D. Costa, F.A. Rodrigues, G. Travieso and P.R.V. Boas, *Advances in Physics*, 2007, **56**, 167.
13. M.E.J. Newman, *Siam Review*, 2003, 45, 167.
14. E. Ravasz and A.-L. Barabasi, *Phys. Rev. E*,. 2003, **67**, 026112.

A LATTICE BOLTZMANN-LATTICE GAS AUTOMATA MODEL FOR ANALYSING THE STRUCTURE AND EFFICIENCY OF DEPOSITED PARTICULATE MATTER ON FIBROUS FILTERS

H. Wang, H. Zhao, Z. Guo and C. Zheng

State Key Laboratory of Coal Combustion, Huazhong University of Science and Technology, Wuhan 430074, China
Email: klinsmannzhb@163.com

1 INTRODUCTION

Fibre filtration, which has the advantage of high collection efficiency of submicron particle, is widely used for air purification at coal-fired power plants, mining engineering, cement industries, work places and life areas. Since the fifties of last century, scientists have been developing mathematical models or carrying out experimental investigations to predict and improve filter performance. However, the filtration process of suspended particles from the airflow is very complicated because of various deposition mechanisms (diffusion, interception, inertial collision, etc) of solid particles as well as the dynamic evolution of filtration efficiency and pressure drop during the filter clogging. In order to construct the optimal fibrous structures with high filtration efficiency (especially for particles of 0.1~1 μm), low pressure drop and long lifetime of the filter, it is essential to obtain the detailed information of a non-steady-state filtration process. Numerical simulation of gas-solid two-phase flow provides a promising approach to investigate the deposition of particulate matter on filter fibres.

Conventional numerical methods, which use Navier-Stokes equations for flow fields and Lagrangian approaches for particle motion, are usually utilized to simulate gas-particle flow in filters. Particular difficulties arise from the non-steady-state and irregular surface cased by deposited particles. These conventional numerical methods thus have to use high-resolution and adaptive grids near the deposit and fibre, on the cost of computational efficiency and precision. The Lattice-Boltzmann (LB) method is an efficient alternative for gas-solid flows with complex and dynamic boundary conditions. The outstanding advantages, compared to the conventional numerical methods, include its simple physical models, inherent parallelism, intrinsic stability and capability to deal with complex and dynamic boundary conditions. Filippova and Hänel[1] first used the LB method for the fluid phase in combination with a Lagrangian approach for the discrete particles to describe non-steady-state filtration. They argued that the LB methods may be the most simple and effective methods to simulate gas-solid flow around such irregular and dynamic geometrical boundaries. Lantermann and Hänel[2] further proposed so-called particle Monte Carlo method to model the motion and trajectory of single particles. These resultant LB-Lagrangian models for gas-solid two-phase flows may lose the ability of parallel computation (which is inherent to the LB methods). Gradon et al.[3,4] used Lattice Gas

Automata (LGA) model for particle movement, deposition and re-entrainment from the deposited structure, where solid particles are constrained to move only on the same regular lattices as the fluid particles. They simulated the filtration process of small particles with a single fibre[3] and the composites of nano- and micro- size hybrid fibres[4], considering the effects of convective diffusion and Brownian diffusion on deposition. However, the LB-LGA model is incapable of quantitatively describing the filtration process due to incorrect or empirical formulation for fluid-particle interaction.

In this paper, we improve the LB-LGA model using an accurate fluid-particle interaction, where the motion probability of a particle to neighbouring nodes due to convective and Brownian diffusion is accurately determined from its actual displacement under consideration of different forces (e.g., Brownian force, drag force). The coupled LB-LGA model is demonstrated for gas-particle flow through parallel cylinders, in such a way that the non-steady-state filtration process is simulated. The simplest filter fibre, a system of parallel cylinders in a laminar flow normal to their axis, is considered and the dynamic evolutions of dendrite-like clusters of particles deposited on a single fibre and deposition efficiency during the clogging process are obtained.

2 NUMERICAL METHODS

2.1 LB model for Flow Field

Gas flow and potential fields are first computed using the classical LB method, where the nine-speed square lattice in a two-dimensional domain (denoted D2Q9) is used. The probability density distribution function $f_i(x,t)$ can be interpreted as the probability of finding some fluid molecules with velocity c_i at time t in the node of the lattice with coordinaters x. Macroscopic density ρ and momentum $\rho\mathbf{u}$ are defined as follows:

$$\rho = \sum_{i=0}^{Q-1} f_i \,, \quad \rho\mathbf{u} = \sum_{i=0}^{Q-1} f_i\mathbf{c}_i \tag{1}$$

The evolution equation of distribution function $f_i(x,t)$ is given by the discretized BGK model:

$$f_i(x + e_i\Delta t, t + \Delta t) - f_i(x,t) = [f_i^{eq}(x,t) - f_i(x,t)]/\tau \tag{2}$$

where τ is the dimensionless relaxation time, and it relates to the fluid viscosity by $\upsilon = c_s^2(2\tau - 1)/2$. The equilibrium distribution function is calculated as follows:

$$f_i^{eq} = \rho\alpha_i[1 + \frac{\mathbf{e}_i \cdot \mathbf{u}}{c_s^2} + \frac{1}{2}(\frac{\mathbf{e}_i \cdot \mathbf{u}}{c_s^2})^2 - \frac{\mathbf{u}^2}{2c_s^2}] \tag{3}$$

where $\mathbf{e}_i$ is the unit vector in direction i (shown in Figure 1); α_i is the weight coefficient related to the model, $\alpha_0 = 4/9$, $\alpha_i = 1/9(i=1,3,5,7)$, $\alpha_i = 1/36(i=2,4,6,8)$; c_s is the local speed of sound, usually equal to $\sqrt{3}/3$. The pressure of a discrete node is calculated as $P = \rho c_s^2$.

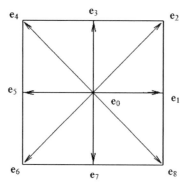

Figure1 *Nine velocities in the D2Q9 Lattice Boltzmann model of a two-dimensional fluid*

2.2 LGA model for Particle Motion

As for the flow of particle-laden fluids, the motion of particles is influenced by not only fluid drag force also random Brownian diffusion. In this paper, we use Brownian force[5] to model the Brownian diffusion. The particle motion is thus described by the Newtonian equation with consideration of drag force F_D and Brownian force F_B:

$$F_p = \frac{du_p}{dt} = F_D + F_B = \frac{u_f - u_p}{\tau_p} + \frac{\varsigma}{Pe}\sqrt{\frac{2d_p^2}{D\Delta t}} \tag{4}$$

where u_p is the particle velocity; τ_p is particle relaxation time scale, $\tau_p = \rho_p d_p^2/(18\mu)$; μ is the dynamic viscosity of gas; ς is the Gaussian random number with average 0 and variance 1; D is the Brownian diffusion coefficient, $D = k_B T/(3\pi\mu d_p)$; d_p is the particle diameter; Pe is the Peclet number, $Pe = Ud_f/D$; U is the average velocity for the stream; d_f is the fibre diameter; k_B is Boltzmann constant; T is temperature, Δt is time-step.

The particle velocity and displacement after the n-th time-step Δt can be explicitly calculated through the integration of Equation (5) over time t:

$$u_p^{n+1} = u_p^n \cdot \exp(-\frac{\Delta t}{\tau_p}) + (u_f + F_B \cdot \tau_p)\cdot(1 - \exp(-\frac{\Delta t}{\tau_p})) \tag{5}$$

$$x_p^{n+1} = x_p^n + (u_p^n - u_f)(1 - \exp(-\frac{\Delta t}{\tau_p})) + u_f \cdot \Delta t + (\Delta t + (1 - \exp(-\frac{\Delta t}{\tau_p})\cdot\tau_p))\cdot F_B \cdot \tau_p \tag{6}$$

where superscript "n" represents the time moment before the n-th time-step Δt, "$n+1$" represents the time moment after Δt. The actual particle displacement Δx ($= x_p^{n+1} - x_p^n$) during the period Δt is thus obtained.

The motion probability of a particle to neighbouring node i is determined by the ratio of its actual displacement on the direction i within Δt and the lattice length of direction i. That is, the probability p_i is proportional to the projection of its displacement Δx on the lattice direction i (see Figure 2):

$$p_i = \max(0, \frac{\Delta x \cdot \mathbf{e}_i}{dx}), (i = 1,3,5,7) \tag{7}$$

where dx is the lattice length. The solid particle located in a lattice node may still stay there or jump to a nearest-neighbour node, depending on the motion probabilities to directions 1, 3, 5, 7:

$$x_p^{n+1} = x_p^n + \mu_1 \mathbf{e}_1 + \mu_3 \mathbf{e}_3 + \mu_5 \mathbf{e}_5 + \mu_7 \mathbf{e}_7 \tag{8}$$

where μ_i is a Boolean variable which is equal to 1 with probability p_i.

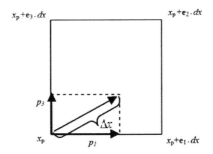

Figure 2 *Particle motion rule in the LGA model*

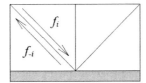

Figure 3 *bounce-back boundary condition in the LB method*

2.3 Treatment of Dynamic Boundary

As the particles deposited, the shape of fiber surface continues to change, resulting in some dendrite-like clusters consisting of deposited particles. Not only the fibre but also clusters serve as particle collectors. Furthermore, the dynamic evolving dendrites affect the flow fields and thus affect the particle deposition. It is necessary to consider the interaction between two-phase flow fields and surface geometry. In the LB methods, the complex boundary conditions can be dealt with some simple rules. It is assumed that any particle is deposited once it touches fiber surface or another deposited particle. Any lattice lying inside a deposited particle is altered from "fluid node" to "boundary node". Once fluid molecules collide with boundary node, they bounce back (Figure 3). It implies that bounce-back boundary condition (BBC) with second-order accuracy, as shown in Figure 3, is employed. The BBC expression is given as:

$$f_{-i}(x_0, t) = f_i(x_0, t) \tag{9}$$

3 RESULTS AND DISCUSSION

The coupled LB-LGA model for gas-solid two-phase flows is used to simulate the deposition of particulate matter on a single filter fibre, where periodic boundary condition of computational domain is used. We mainly focus on the dynamic evolution of deposition patterns[6] and collection efficiency[7] when various deposition mechanisms (diffusion, interception, inertial collision) are dominant.

3.1 Dynamic Evolution of Deposition Patterns Dominated by Various Deposition Mechanisms

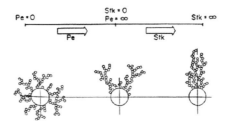

Figure 4 *Deposition patterns dominated by various deposition mechanisms[9]*

Kanaoka *et al.*[8] presented the relationship between deposition patterns and filtration condition (in terms of the non-dimensional filtration parameters: Peclet number *Pe* and Stokes number *St*), as shown in Figure 4. The three conditions (1. *Pe*=0; 2. *Pe*=∞ and *St*=0, 3. *St*=∞) are capable of representing Brownian diffusion, interception and inertial collision mechanisms, respectively.[9] Pure Brownian diffusion leads to an isotropic distribution of particles around the fibre with relatively open pore structures, and the interception mechanism results in two striking dendrites located on 45° and 135° of cylinder surface and growing up along the opposite direction of flow steam, and the inertial collision mechanism makes the particles deposit on the windward of the fibre. These deposition patterns have been observed experimentally. The LB-LGA model provides the detailed knowledge on the dynamic evolution of dendrites, as shown in Figure 5. Good agreement in the shape of dendrites with those reported in literature was obtained. It is proven that the LB-LGA model is able to capture the non-steady-state filtration process correctly.

Factually, when Brownian diffusion is dominant (that is, *Pe* is very small) the Brownian force F_B is far larger than the drag force (Eq. 4), resulting in a very stochastic trajectory of particles (see Figure 6a). Some particles can randomly move to the leeward and deposited due to collision with the fibre. If *Pe* becomes very large and *St* is very small, Brownian force of particles is negliable and particles fully follow the steamline of fluid (Figure 6b), resulting in particle deposition in two special positions of windward cylinder. With regard to dominant inertial collision mechanism, very large *St* means particles have very high inertia and can not deviate from their original directions, leading to particles' head-on collision on fiber surface (see Figure 6c), although the fluid will move around the obstacle fibre.

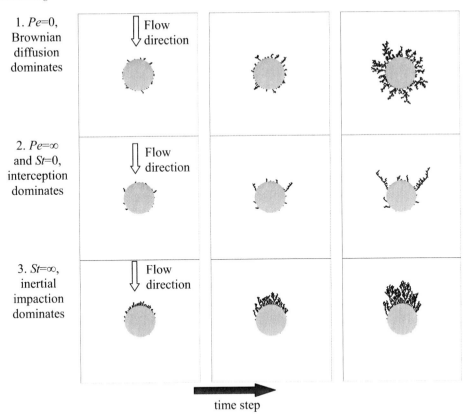

Figure 5 Dynamic *evolution of deposition patternsobtained using the LB-LGA model*

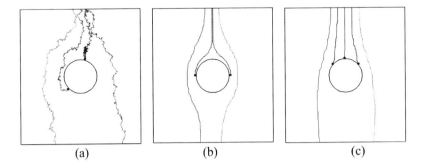

Figure 6 *Particle trajectories with different mechanisms*

3.2 Capture Efficiency of Dust-loaded Fibres

The growth of branch cluster structure results in the expansion of capture range and increasing caprute efficiency of fibres. Kasper *et al.*[7] proposed an expirical equation to characterisize the increasing in capture efficiency and the mass of loaded particles. Our simulation results are in excellent agreements with the preditions of Kasper *et al.,*[7] which further demonstartes the ability of the LB-LGA model for fibre filteration.

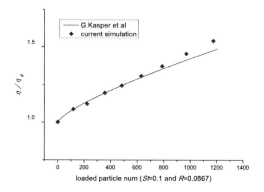

Figure7 *Capture efficiency vs loaded particles number*

4 CONCLUSIONS

The LB-LGA model offers a useful tool for modelling the unsteady filtration process, including the dynamic evolution of the branch cluster structure and capture efficiency along with particle loading. In the LB-LGA model, the interaction between gas-solid flow and complex and unsteady geometric boundary is readily implemented. Furthermore, the LB-LGA model uses the same lattice grids, which still conserves the advantages of LB methods, such as inherent parallelism. The LB-LGA model is capable of quantitatively analysing the filtration process, due to correct consideration of fluid-particle interactions. The numerical results are in good agreement with both previous theoretical predictions and experimental observations. The reliable predictions will help construct the optimal fibrous filter structures, and provide basic model parameters and constitutive equations for macroscopic modelling of penetration and pressure drop.

References

1   O. Filippova and D. Hanel, *Computers & Fluids*, 1997. **26**(7), 697.
2   U. Lantermann and D. Hanel, *Computers & Fluids*, 2007. **36**(2), 407.
3   R. Przekop, A. Moskal and L. Gradoń, *Journal of Aerosol Science*, 2003. **34**, 133.
4   R. Przekop and L. Gradoń, *Aerosol Science and Technology*, 2008. **42**, 483.
5   H.N. Unni and C. Yang, *Journal of Colloid and Interface Science*, 2005. **291**, 28.
6   S. Hosseini and H. Tafreshi, *Separation and Purification Technology*, 2010, **74**,160.
7   G. Kasper, S. Schollmeier, J. Meyer and J. Hoferer, *Journal of Aerosol Science*, 2009. **40**, 993.
8   C. Kanaoka, H. Emi, and T. Myojo, *Journal of Aerosol Science*, 1980. **11**(4), 377.
9   G. Kasper, S. Schollmeier and J. Meyer, *Journal of Aerosol Science*, 2010, **41**, 207.

NUMERICAL SIMULATION OF A PIPE LEAKAGE PROBLEM WITH COHESIVE PARTICLES

X. Cui, J. Li, A. Chan and D. Chapman

School of Civil Engineering, University of Birmingham, Birmingham, B15 2TT, UK

1 INTRODUCTION

Leakage from underground pipes is a common problem with buried services and results in, for example, expensive water losses, potential soil contamination and subsidence. One way of visualising a leak in a buried pipe is by having a fluid passing upwards through a narrow slot ('leak') and impacting the soil above. This could result in an idealised model of a particle bed subject to a localised fluid flow. Soil particles surrounding the leak could be washed away by the injecting fluid, generating a subsurface cavity. This cavity may develop gradually, and expose the buried infrastructure to the danger of collapse, or lead surface subsidence. Due to the importance of maintaining the underground environment and associated safety issues, there is a need to understand how soils respond to local leakage, especially for cohesive soils, which are commonly faced in the field.

Compared with field studies and laboratory experiments, numerical simulations provide a more flexible and efficient way in 'visualising' the material behaviour in response to a leaking fluid without any field or sample disturbances. As a real challenge is raised attributed to complicated interactions between the soil and the leaking fluid in the vicinity of the 'leak', the capability of tracing the fluid-particle interactions at a particle level is desired around the 'leak'. A numerical technique coupling the Discrete Element Method (DEM) and the Lattice Boltzmann Method (LBM)[1,2] is adopted in this study.

2 METHODOLOGY

2.1 Soft-sphere DEM

The coupled DEM-LBM technique is regarded as a powerful and efficient tool in simulating fluid-particle systems in small-scale modelling. The DEM analyses the granular soil on a particle level. In DEM, granular material is considered as an assembly of separate particles. When the soft-sphere approach is used,[3] a slight overlap is allowed between two particles in contact, and the interactions are viewed as a dynamic process in which contact forces accumulate or dissipate over time. Contact forces can be subsequently obtained through the deformation history at the contact. The motion of a single particle is governed by Newton's second law in the form of the following dynamic equations,

$$m\frac{d^2\vec{x}}{dt^2} = \vec{F}_c + \vec{F}_b + \vec{F}_h \tag{1}$$

$$I\frac{d\vec{\omega}}{dt} = \vec{T}_c + \vec{T}_h \tag{2}$$

where $\vec{F}_c$ denotes the total contact force, calculated by summing up the contact forces acting on one particle. $\vec{T}_c$ indicates the total torque generated by the contact force. $\vec{F}_b$ represents the body force, namely, the submerged gravity in this work. $\vec{F}_h$ and $\vec{T}_h$ refer to the force and torque applied by the flowing fluid, respectively. Their values are obtained through the fluid-particle two-way coupling (see Sec. 2.3).

The contact force calculations from the Birmingham DEM code originally developed by Thornton[4,5,6] were directly incorporated into the code developed for this study. The auto-adhesive elastic contact model is used to simulate the inter-particle cohesion. In this contact model a 'pull-off force' is adopted, which represents the maximum tensile force required to break a contact and given as,

$$P_c = 3\gamma\pi R^* \quad \text{with} \quad \frac{1}{R^*} = \frac{1}{R_A} + \frac{1}{R_B} \tag{3}$$

where γ is the surface energy of each solid particle, and R_A and R_B are the radii of the two particles in contact. Accordingly, in order to examine the effects of inter-particle cohesion in this study, values of γ were selected according to the following relationship,

$$\overline{P}_c = K\overline{mg} \tag{4}$$

where K indicates the ratio of the average bond strength to the average particle weight.

With the accelerations computed from Eqns. (1) and (2) for each particle at each time step, the particle velocities are integrated using a central difference scheme, and the location of each particle is updated for calculations of the next cycle.

2.2 LBM with Large Eddy Simulation (LES)

The LBM is used as an effective alternative to conventional macroscopic methods for fluid flow simulations. It is a time-stepping procedure, based on microscopic kinetic models. In the classic LBM, the fluid domain is divided into a rectangular or cubic lattice with uniform spacing. Fluid is viewed as packets of micro-particles residing on the lattice nodes. During each time step, particles are allowed either to remain in the same node, or to travel to their adjacent nodes with corresponding discrete velocities $\vec{e}_i (i = 0,...,n)$. In order to simulate fluid with high Reynolds numbers, the Large Eddy Simulation (LES) technique has been implemented with LBM.[7] The governing LBM equation in a turbulence filtered form is given as follows,

$$\widetilde{f}_i(\vec{x} + \vec{e}_i \Delta t_{LBM}, t + \Delta t_{LBM}) = \widetilde{f}_i(\vec{x}, t) - \frac{1}{\tau_{total}} [\widetilde{f}_i(\vec{x}, t) - \widetilde{f}_i^{eq}(\vec{x}, t)] \quad (i = 0, ..., n) \tag{5}$$

where $\widetilde{f}_i$ are filtered density distribution functions, and $\widetilde{f}_i^{eq}$ are a set of filtered equilibrium distribution functions related to the prescribed discrete velocities and the macroscopic variables, including fluid density and velocity. The effect of the unresolved scale, which is principally responsible for energy dissipation through viscous forces, is considered by a relaxation time τ_{total}, taking into account the effect of turbulence viscosity.

2.3 Coupling of Solid and Fluid Phases

The Immersed Moving Boundary (IMB) scheme is adopted for fluid-particle inter-phase treatment. At each lattice node, the LBM equation is modified using a weighting function which is associated with the volume fraction of the corresponding nodal cell (the yellow square in Figure 1) covered by a moving solid particle. In such a way, the local fluid flow influenced by the presence of solid particles can be taken into account. On the other hand, smooth and accurate hydrodynamic force and torque acting on the solid particle can also be obtained. The detailed implementation of this coupling scheme is illustrated in Noble and Torczynski (1998).[8]

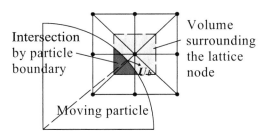

Figure 1 *Volume fraction of a nodal cell covered by a moving solid particle*

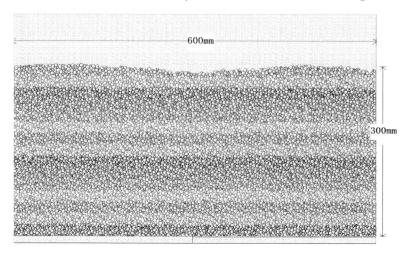

Figure 2 *Numerical setup*

3 NUMERICAL SETUP AND RESULTS

A two-dimensional numerical simulation was carried out using the coupled DEM-LBM technique. A densely packed bed consisting of 9,997 circular particles was settled in a rectangular container (Figure 2). A particle size range from 3.0mm to 6.0 mm was used, which was uniformly distributed. The bed had dimensions of 600 mm in length and 300 mm in height. A small orifice with width 3.0 mm was located in the middle of the bed base, and this connected to a horizontal fluid pipe underneath the container, allowing the fluid to enter the bed at the base.

Table 1 *Various values of surface energy adopted in the tests*

K	$\gamma\,(\text{J/m}^2)$
0	0
1	0.0765
5	0.3830
10	0.7650
20	1.5300

In order to investigate the effects of cohesive particles on the bed behaviour subject to a localised fluid injection, simulations have been conducted with the controlled surface energies listed in Table 1 (see also Eqns. (3) & (4)).

In all the cases, a same initial particle configuration is used, and the inlet orifice velocity is kept constant at 3.0 m/s. With such a high velocity, all the cases eventually undergo a 'blow-out' failure (Figure 3) in which a cavity forms and continuously develops until the particle bed is ruptured up to its top surface. A typical snapshot of the particle configuration after the cavity forms is provided in Figure 3. For the bed with non-cohesive particles ($K = 0$), the bed is raised smoothly without any obvious cracks or fractures. The cavity is almost round at the top. However, in the case of $K = 5$, some fine cracks appear at both the top surfaces of the cavity and the bed, and the cavity has a more irregular shape. When the K value is increased to 20, very large fractures are observed, and the soil particles aggregate, behaving as larger-sized blocks.

4 CONCLUSION

A DEM-LBM simulation was conducted on a pipe leakage model with different surface energy values. Results show the existence of inter-particle cohesion has a significant effect on the behaviour of the particle bed when subject to a locally injected fluid, including cavity shapes, as well as particle behaviour by 'gluing' them together and generating larger-sized blocks. In addition, cracks and fractures are observed at both the top surfaces of the cavity and the bed with relatively large surface energy values.

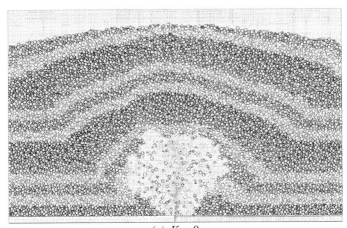

(a) $K = 0$

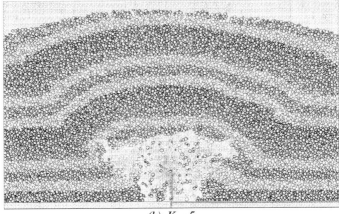

(b) $K = 5$

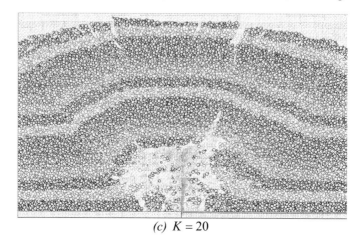

(c) K = 20

Figure 3 *Examples of particle configurations with different K values*

References

1 B.K. Cook, D.R. Noble, and J.R. Williams, *Engineering Computations*, 2004, **21**(2-4), 151.
2 Y.T. Feng, K. Han, and D.R.J. Owen, International Journal for Numerical Methods in Engineering, 2007, **72**(9), 1111.
3 N.G. Deen, M. Van Sint Annaland, M.A. Van der Hoef, and J.A.M. Kuipers, Chemical Engineering Science, 2007, **62**(1-2), 28.
4 C. Thornton, Journal of Physics D (Applied Physics), 1991, **24**(11), 1942.
5 C. Thornton, and K.K. Yin, Powder Technology, 1991, **65**(1-3), 153.
6 C. Thornton, and Z. Ning, Powder Technology, 1998, **99**(2), 154.
7 K. Han, Y.T. Feng, and D.R.J. Owen, Computers and Structures, 2007, **85**(11-14), 1080.
8 D. Noble, and J. Torczynski, Int. J. Mod. Phys. C, 1998, **9**, 1189.

CFD INVESTIGATION OF PLUG FLOW DENSE PHASE PNEUMATIC CONVEYING WITH A STEP IN PIPELINE BORE

D. McGlinchey, A. Cowell and R. Crowe

School of Engineering and Computing, Glasgow Caledonian University, Glasgow, Scotland, UK

1 INTRODUCTION

The use of stepped pipeline in pneumatic conveying systems can be advantageous in reducing pipeline erosion, product degradation and flow resistance.[1,2] This is due to the reduction in conveying gas velocity following an increase in pipe cross section. The benefits of this approach are obvious for high pressure dilute phase conveying systems, where a large portion of the available energy is used to overcome the friction experienced by the gas phase. It may also be appropriate for dense phase systems, for example, to prevent plugs reaching high velocity with the potential to damage system components and supports. The detail of the flow behaviour of dense phase plugs in passing through a step does not seem to be covered in the literature and some initial work is reported in this paper.

2 METHOD AND RESULTS

The flow behaviour of a 500 mm long plug of material, particle size 25 μm, particle density 2500 kg/m^3 passing through: an abrupt step (sudden expansion) and a gradual expansions was investigated. A 3000 mm section of conveying line with a step from 75 mm bore to 100 mm bore was modelled and meshed in 3D in Gambit 2.3.

2.1 Mathematical Model

There are many approaches to model pneumatic conveying of particulate solids; including 1D models,[3] combined CFD-DEM[4,5,6] and a two fluid model.[7] The model chosen for this work was the Euler-Euler based model with the solids phase represented by the granular kinetics formulation as implemented in the code by Fluent 6.3. This approach has been used by workers studying fluidised beds[8] and is appropriate for modelling fluidised bed dense phase pneumatic conveying and has been used successfully in the past for a variety of pneumatic conveying studies.[9,10,11]

The following assumptions were made: no mass transfer between the phases or source terms; lift forces are due mainly to velocity gradients in the gas phase and were assumed negligible for the small particle size in this case; the virtual mass force has been neglected due to density differences between the solids and gas phases. The simplest of the multi-phase turbulence models, an extension of the single-phase k-ε model, provided in Fluent was used in this study. The model constants were left at default values. The solution method employed for the Eulerian multi-phase calculations was the phase coupled SIMPLE algorithm. A density based solver was used to solve for the velocity components of all phases simultaneously, a pressure correction equation was built based on total volume continuity, then the pressure and velocities were corrected to satisfy continuity.

2.2 Boundary Conditions

A 'velocity inlet' was defined with a gas velocity of 3 m/s and no solids introduced. The outlet was set to a 'pressure outlet'. The gas phase was set to obey the no slip condition at the pipe wall with no special wall treatment. The choice of boundary condition for the solids phase is less certain, the choice being between; no slip, partial slip and free slip. Direct observation from experiments with dense phase flow of fine particulate material through a glass section of pipeline shows that there is slip between the moving material and the pipe wall. Experimental and numerical work by Lain and Sommerfeld[12] in the dilute phase region has identified the importance of wall roughness in the consideration of particle motion in pneumatic conveying lines and Li *et al.*[13] reported a study on the effect of the wall boundary condition in the simulation of bubbling fluidised beds highlighting the importance of this parameter. However, CFD analysis of fluidised beds by Chen *et al.*[14] suggests that realistic behaviour may be obtained by applying the free slip condition for their case. As a working hypothesis this condition was also applied in this study, however, this may require to be changed in light of experimental evidence. A full bore 500 mm long plug of material, volume fraction 0.6, velocity 3 m/s, was 'patched' at a location 500 mm from the inlet. A transient simulation with a fixed time step of 0.0001 seconds was performed. Monitors were positioned at locations along the length of the pipe and in particular at the step to record volume fraction at 0.001 second intervals and all data were saved at 0.1 second intervals for a total run time of 1 second. The models and boundary conditions used in the simulations are summarised in Table 1.

2.3 Results and Discussion

The results from this study of the effect of enlargement geometry were mainly interpreted from contour plots of solids volume fraction and velocity vector plots of solids velocity, and are discussed in terms of the implications for pneumatic conveying system operation and design. The solids volume fraction was monitored at the centre of the pipe cross-section location at the start of the enlargement and the values of solids volume fraction against flow time for the sudden expansion and the gradual expansion over 300 mm is shown in Figure 1.

It can be seen from Figure 1 that the solids volume fraction at the centre of the pipe at the location of the enlargement is clearly different for the sudden or abrupt enlargement than for the gradual enlargement case. In the sudden expansion case, the solids volume fraction value starts to rise from zero at a flow time of approximately 0.16 seconds. This is consistent with the distance from the front of the plug to the monitoring point being 0.5 m and the plug initial velocity being set at 3 m/s. The solids volume fraction then rises to a value of 0.32 at 0.2 seconds, falls to 0.11 at 0.24 seconds before rising again to a maximum value of 0.49 at a flow time of 0.27 seconds. There is a further fall and rise to a peak of 0.41 at a flow time of 0.34 seconds before the solids volume fraction decreases back to a near zero value at 0.47 seconds. The total time taken for the plug to pass the monitoring point is then seen to be 0.31 seconds. If the plug had retained its original shape and velocity, the plug would have passed the monitoring point in 0.167 seconds. The initial relatively rapid rise over 0.04 seconds and the much slower decline over 0.13 seconds as the plug leaves the monitoring point is consistent with a plug shape which has a relatively flat front and a long tail.

The variation of solids volume fraction in the plug at the monitoring point is an interesting phenomenon that does not appear to have been reported previously. The times at initial rise and return to zero value of solids volume fraction for the gradual expansion case are the same as the abrupt expansion case; however, the pattern of solids volume fraction for the gradual expansion is very different from the abrupt expansion case. There is a rapid rise at the same time of around 0.16 seconds, however, there follows a fluctuating pattern of solids volume fraction between values of approximately 0.25 and 0.05 over a flow time of 0.14 seconds with a tail of decreasing solids volume fraction values occurring over a further 0.2 seconds before returning to zero.

Table 1 *Simulation models and boundary conditions*

Parameter	Value
Granular viscosity	Gidaspow[15]
Granular bulk viscosity	Lun et al.[16]
Frictional viscosity	Schaeffer [17]
Angle of internal friction	30°
Granular temperature	Algebraic
Drag law	Gidaspow[15]
Particle Size	2.5×10^{-5} m
Particle Density	2500 kg/m3
Coefficient of restitution for particle–particle collisions	0.9
Inlet boundary condition	Velocity inlet
Outlet boundary condition	Pressure outlet
Wall boundary condition	No slip for air, Free slip for solids
Pipe bore initial	0.075 m
Pipe bore final	0.1 m
Total pipe length	3 m
Initial plug length (full bore)	0.5 m
Initial volume fraction of solid phase	0.6
Initial solids velocity	3 m/s
Inlet gas velocity	5 m/s
Viscous model	Standard k – ε model
Wall treatment	Standard
Pressure Velocity Coupling	Phase coupled SIMPLE
Discretization	Fist order upwind
Time step	$1 \times 10{-}4$ s

NOTE: For details see Fluent User Giude[18]

From Figure 2 it can be seen that again there is a clear difference between the abrupt and gradual expansion cases. Material can be seen to arrive at the outlet much earlier than would have been the case if the plug had remained coherent in either the abrupt of gradual expansion case. The early arrival of material at the outlet is particularly marked in the gradual expansion case where solids appear at the outlet before 0.2 s flow time. Study of volume fraction contour and solids velocity vector plots show that the plug in the gradual expansion case has material taken from the top surface of the plug by high velocity air currents and arrives at the outlet with low solids volume fraction and velocities in the region of 25 m/s. The same general behaviour can be seen for the abrupt expansion case, however, the speed of travel is reduced. The variation in solids volume fraction with time shows that the initial plug is no longer travelling in a coherent manner.

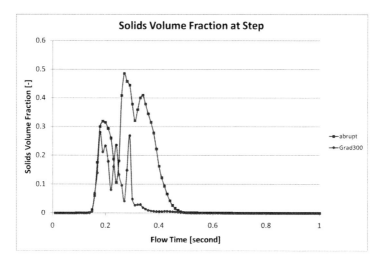

Figure 1   *Solids volume fraction vs. flow time for the sudden expansion (abrupt) and the gradual expansion over 300 mm (Grad300) at the location of the step*

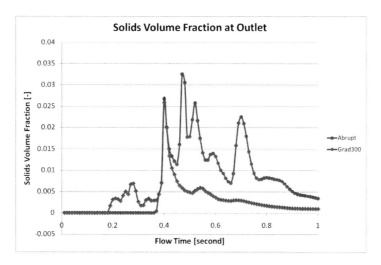

Figure 2   *Solids volume fraction vs. flow time for the sudden expansion (abrupt) and the gradual expansion over 300 mm (Grad300) at the location of the pipeline outlet*

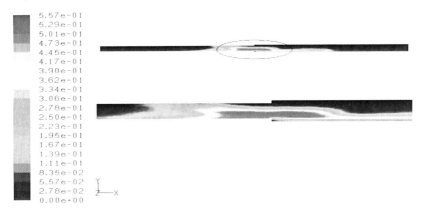

Figure 3 *Abrupt expansion solids volume fraction section view at 0.3 s*

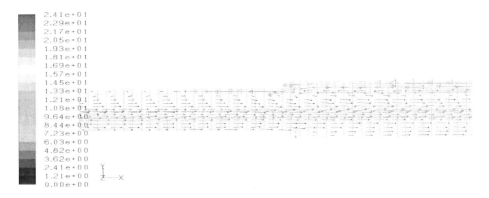

Figure 4 *Abrupt expansion solids velocity vector section view at 0.3 s*

Study of volume fraction contour (Figure 3) and velocity vector (Figure 4) plots for the abrupt step case show that the solids can be seen to 'flow over' this step; that is, there is a low value of solids volume fraction at the location of the step itself. The region following the step has low values of solids volume fraction close to the lower wall, increasing value to the centre of the plug then decreasing again to the top of the pipe. The plug moves further past the step with an overall lower value of solids volume fraction, If this is a realistic picture, then the material is likely to increase in aeration or at least remain aerated following transport through an abrupt enlargement which is beneficial to this mode of conveying. In contrast, the flow pattern seen in the gradual expansion case shows a plug which has 'collapsed' and elongated along the length of the expansion section. High values of solids volume fraction are seen close to the lower wall. This could indicate that for some materials there is potential for the material to de-aerate with a resulting large increase in resistance to flow; the consequence of which may be a pressure surge or a potential blockage in the line. Examination of air and solids velocity plots revealed a recirculation zone seen at the location of the abrupt step as may be expected from single phase behaviour. The solids velocities at the top of the pipe were over 20 m/s. The volume fraction is very low, however, at these velocities there is a potential for pipeline wear. There is also a change in direction of the particles out of the plane. This suggests that particles are impacting the wall at the step which may result in wear or in particle degradation. A large recirculation zone was seen in the gradual step and is not expected in single phase and is therefore thought to be the result of the presence of solids in the pipe. Again there were regions with high velocity particles, over 25 m/s, at low solids concentration which could produce pipeline wear or product degradation.

3 CONCLUSIONS

The effect of the geometry of an expansion in pipeline bore on the flow behaviour of a dense phase 'plug' of fine particulate material being conveyed pneumatically through a pipeline has been investigated using a 3D CFD model. The solids flow behaviour was interpreted using plots of solids volume fraction and solids velocity vector plots. These indicate that an abrupt expansion may result in a plug which has passed through the step having increased aeration, whereas for the gradual expansion case the plug was seen to collapse and elongate with a potential for de-aeration and pipeline blockage. Both the abrupt and gradual expansion geometries resulted in recirculation zones of solids high velocity but low solids concentration which could be potential sites for pipeline wear or particle degradation. The recirculation zone in the gradual expansion case was shown to be caused by the presence of the solids.

References

1. D. Mills, *Pneumatic Conveying Design Guide, 2nd Edition*, 2004, Butterworth-Heinemann, ISBN: 0750654716
2. G.E. Klinzing, R.D. Marcus, F. Rizk, L.S. Leung, *Pneumatic Conveying of Solids; A Theoretical and Practical Approach, 3rd Ed.*, 2010, Springer, London.
3. D.J. Mason, P. Marjanovic, A. Levy, *Powder Technol*, 1998, **95**, 7.
4. Y. Tsuji, T. Tanaka, T. Ishida, *Powder Technol*, 1992, **71**, 239
5. J. Xiang, and D. McGlinchey, , *Granul Matter*, 2004, **v6, n 2-3**, 167
6. J. Xiang, D. McGlinchey, J-P. Latham, *Granul Matter*, 2010, **12**, 345
7. D.J. Mason, A. Levy, *Int J Multiphas Flow*, 2001, **27**, 415
8. L. Cammarata, P. Lettieri, G. Micale, D. Colman, *Int J Chem React Eng*, 2003, **1**, 1
9. C. Ratnayaka, M.C. Melaaen, B.K. Datta, *Proceedings of 8th International Conference on Bulk Materials Storage, Handling and Transportation, Wollongong, Australia*, 2004, 388
10. D. McGlinchey, A. Cowell, *Bulk Solids & Powder Sci & Technol*, 2008, **3**, 109
11. C. Zhang, Y. Zhang, D. McGlinchey, Y. Du, X. Wei, L. Ma, C. Guan, *Powder Technol*, 2010, **204**, 268
12. S. Lain, M. Sommerfeld, *Powder Technol*, 2008, **184**, 76
13. T. Li, J. Grace, X. Bi, *Powder Technol*, 2010, **203**, 447
14. X-Z. Chen, D-P. Shi, X. Gao, Z-H. Luo, *Powder Technol*, 2011, **205**, 276

15. D. Gidaspow, *Multiphase Flow and Fluidization*. Academic Press, Boston, 1994.
16. C.K.K. Lun, S.B. Savage, D.J. Jeffery, N. Chepurnity, , *J Fluid Mech*, 1984, **140**., 223
17. D. G. Schaeffer, *J Differ Equations.,* 1987, **66**, 19
18. *Fluent 6.3 Users Guide*, ANSYS Inc.

CFD SIMULATION OF FLUID FLOW AND HEAT TRANSFER IN A COUNTER-CURRENT REACTOR SYSTEM UNDER SUPERCRITICAL CONDITIONS

C. Y. Ma[1], T. Mahmud[1], X.Z. Wang[1], C. Tighe[2] and J.A. Darr[2]

[1] Institute of Particle Science and Engineering, School of Process, Environmental and Materials Engineering, University of Leeds, Leeds LS2 9JT, UK
[2] Department of Chemistry, University College London, London WC1H 0AJ, UK

1   INTRODUCTION

Hydrothermal treatments with supercritical water offer great advantages in the manufacture of homometallic or heterometallic oxide nano-materials.[1] Continuous hydrothermal flow synthesis (CHFS) method was developed to overcome the limitations of hydrothermal batch reactions.[2] CHFS is green, and easy to control in manufacture of functional nano-materials.[3, 4] In scaling-up a process from laboratory to pilot scale, it is anticipated that there are some 'scaling-up' factors that are difficult to study purely using experiments. Process modelling using computational fluid dynamics (CFD) is a useful tool to study the effects of these factors at different scales and evaluate alternative designs. The application of CFD modelling to supercritical water hydrothermal synthesis systems is limited in literature. Lester et al.[5] used CFD technique to simulate the velocity distribution of a nozzle reactor using methanol and sucrose to represent supercritical water and metal salt solution. Aimable et al.[6] carried out numerical simulation in the mixing zone of an X-shaped reactor and obtained the temperature and velocity distributions without reaction. Some researchers investigated tubular heat exchangers using supercritical fluids in various applications such as supercritical water nuclear reactors[7-10] and supercritical CO_2 air-conditioners and heat pumps.[11-14] However, most of the studies[7-13] applied a constant heat flux to the outer wall of a vertical or horizontal circular tube. Cheng et al.[15] found that the heat transfer coefficients from different correlations varied significantly and used CFD to understand heat transfer behaviours in supercritical water flow. The accuracy and validity of some semi-empirical correlations for heat transfer coefficients were compared with measurements at supercritical conditions in an annular flow with a constant heat flux.[16]

In this paper, CFD technique was applied to simulate the processes in a CHFS system.[3, 4, 17] The predictions of flow and temperature profiles in the reactor and heat exchanger were carried out using the Fluent package[18] with a k - ε turbulence model. The heat transfer coefficients in the heat exchanger were estimated using simulated and experimental results. The predicted results were compared with experimental data.

2 MODELLING PROCEDURE AND DETAILS

2.1 Hydrodynamic Models and Thermodynamic Properties

The fluid flow prediction is based on the numerical solution of the three-dimensional, Reynolds-averaged continuity and Navier-Stokes equations, together with the commonly used k - ε turbulence model. The temperature variation was obtained by solving the total enthalpy equation for a single-phase, multi-component system.

In this study, the water properties obtained from the 1995 IAPWS formulation[19] were piece-wise curve-fitted over several temperature ranges at 24.1 MPa. The obtained equations can calculate thermodynamic properties efficiently and accurately.

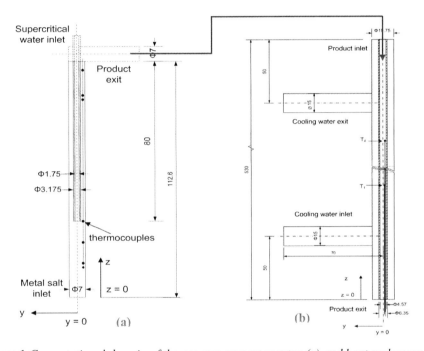

Figure 1 *Computational domain of the counter-current reactor (a) and heat exchanger (b).*

2.2 A Counter-Current Reactor CHFS System

Figure 1a shows a diagram of the counter-current reactor used in a CHFS system.[3, 4, 17] The reactor consists of an inner tube and an outer tube. The superheated water flows downwards in the inner tube to mix with the precursor stream flowing upwards (Figure 1a). The generated product stream flows upwards to leave the reactor and enter a tubular heat exchanger. The precursor stream is mimicked by deionised water, hence the product being hot pressurised water. Seven thermocouples were inserted into the reactor at locations along the z direction (Figure 1a). The estimated tip position variations are within 2 mm across the tube cross-section (x-y plane).

A schematic diagram of the heat exchanger is shown in Figure 1b, which consists of an inner tube located within an outer tube. The hot pressurised water flows downwards through the inner tube while cooling water flows upwards through the shell side. Thermocouples were fed up through the inner tube from the product exit (Figure 1b) to measure the product temperatures at four locations along the z axis. The maximum deviation of the thermocouple position in the radial direction (x-y plane) was < 2 mm.

3 COMPUTATIONAL DETAILS

3.1 Counter-Current Reactor and Tubular Heat Exchanger

Figure 1a shows the computational domain of the reactor which was meshed using GAMBIT.[18] The inlet temperatures of supercritical water and precursor were 400 and 15°C, respectively, at a flowrate of 20 mL/min and a pressure of 24.1 MPa. The computational domain of the heat exchanger (Figure 1b), has three subdomains: inner tube, the wall between the product and cooling water, outer tube and the inlet and exit arms, which were discretised using GAMBIT.[18] The inlet temperatures of product and cooling water were 281 and 10°C, respectively, with the corresponding flowrates being 40 mL/min and 1.5 L/min, and the operating pressures 24.1 and 0.6 MPa.

3.2 Solution Method

The governing equations together with the k and ε equations are solved using Fluent[18] to obtain flow and heat transfer profiles in the counter-current reactor system including the reactor and heat exchanger. Standard SIMPLE pressure-velocity coupling was used with a second order upwind scheme being employed for the discretisation of the convection terms. Adiabatic boundary condition was used due to the insulation of the outer wall of the reactor and heat exchanger. The mass inlet flow mode was used to calculate the inlet velocities for the reactor and heat exchanger. A fully developed outlet flow condition was used for their corresponding exit boundaries. Constant inlet temperatures for the inlet fluids were specified. A turbulent intensity of 10% and the corresponding hydrodynamic diameters were used for the inlet conditions of turbulence. Standard non-slip wall boundary conditions were applied with the standard turbulent wall function. Independence tests for mesh size and convergence tolerance were carried out.

4 RESULTS AND DISCUSSION

4.1 Counter-Current Reactor

Figure 2 shows the contours of velocity and temperature distributions around the exit region of supercritical water, which penetrated into the up-coming precursor stream to form a recirculation zone surrounding the supercritical water jet. This enhanced the mixing between the supercritical water and precursor streams. The mixture then flew upwards through the annual section to the product exit. The predicted and measured temperatures along the z direction were compared in Figure 3. As the locations of the thermocouple tips can vary within 2 mm on the x-y plane in the annual section, i.e. y = 1.5875 ~ 3.5 mm, the predicted results at y = 1.7 and 2.9 mm were plotted in Figure 3. It can be seen that the predicted results are in good agreement with measurements.

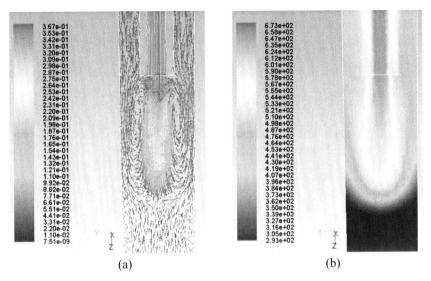

(a)   (b)

Figure 2 *Velocity vectors (a) and temperature contour (b) in the supercritical water exit region of the counter-current reactor*

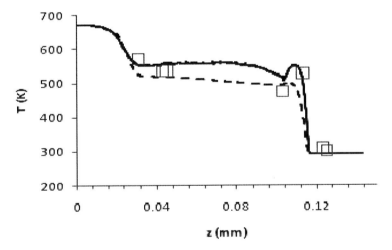

Figure 3 *Predicted (solid line: y = 1.7mm; dash line: y = 2.9mm) and measured (squares) temperatures along the z direction.*

4.2 Tubular Heat Exchanger

Simulated contours of velocity and temperature around the product inlet and cooling water outlet regions are shown in Figure 4. Due to the dead end of the outer tube, a stagnant zone above the exit tube was formed. A recirculation zone was also generated when the cooling water left the outer tube and entered into the exit. As a result, the cooling water temperature in the stagnant zone (Figure 4b) is higher than that in the exit tube. To minimise this region, hence improving heat exchanger performance, the heat exchanger needs to be redesigned to move the exit tube close to the product flow inlet. Similar stagnant zone of cooling water was formed below the cooling water inlet but with little influence on the heat exchanger performance due to low temperatures in this region.

Figure 5a shows the measured and predicted temperatures of the product and cooling water along the z direction. Due to the 2 mm deviation of thermocouple tips, predicted temperatures along the z direction are plotted at $y = 0.0$ and 1.2 mm. Up to around $z \approx 200$ mm, predicted temperature variations between $y = 0.0$ and 1.2 mm were very small, whilst prediction and measurements at $y = 1.2$ mm for $z > 200$ mm were in good agreement, which indicated that the actual thermocouples tips were not located at the axis ($y = 0.0$ mm) instead of $y = 1.2$ mm. The product temperature was reduced from 281 to 100°C within 3 s from its inlet (Figure 5a), corresponding to a cooling rate of 60°C/s, hence providing quenching effect for high quality nano-materials. The simulated and measured cooling water temperatures along the z direction are also in good agreement (Figure 5a).

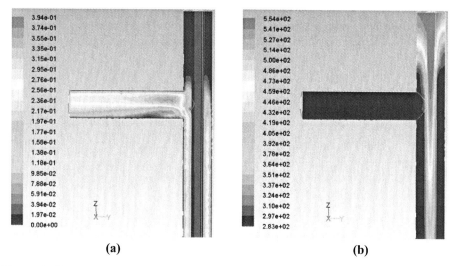

(a)   (b)

Figure 4 *Contour of velocity (a) and temperature (b) in the top part of the heat exchanger.*

Figure 5b shows the heat transfer coefficients estimated from the predicted and measured temperatures along the z direction. The heat transfer coefficients along the z direction were calculated from predicted temperatures using the same method in a previous paper.[20] Good agreements were achieved between the prediction and measurements.

5 CONCLUSION

In this study, CFD modelling was used to predict the flow field and heat transfer profiles in a counter-current reactor and a tubular heat exchanger of a CHFS system. The predicted temperatures along the length were compared with the experimental data. In the heat exchanger, the simulated and measured temperatures along the length were in good agreement. The product temperatures were rapidly cooled down from 281 to 100°C at a cooling rate of ~60°C/s, which effectively discouraged further reaction, hence helping produce high quality nano-particles. The predicted heat transfer coefficients were also in good agreement with measurements. The research demonstrated that CFD modelling provides a powerful tool to predict flow, mixing and heat transfer in counter-current reactors and heat exchangers of CHFS systems, and for the design optimisation and scaling-up of such systems.

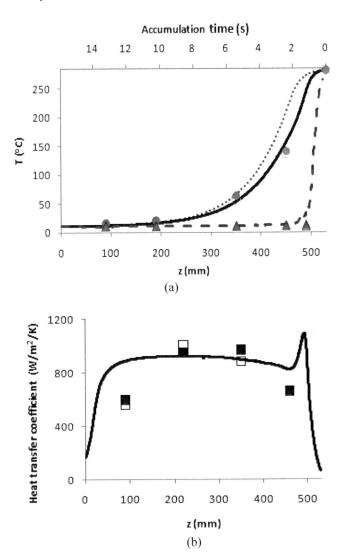

(a)

(b)

Figure 5 *(a) Predicted and measured temperatures along the z direction (product: dot line – prediction at y = 0 mm, solid line – prediction at y = 1.2 mm, circles – measurements; cooling water: dash line – prediction, triangles - measurements). (b) Heat transfer coefficients from simulation (line) and measurements (squares).*

Acknowledgements

Financial support from the UK Engineering and Physical Sciences Research Council (EPSRC) for the EngNano project (EP/E040624/1, EP/E040551/1) is acknowledged. We would also like to thank the industrial collaborators including Johnson Matthey PLC, Corin Group PLC, Resource Efficiency Knowledge Transfer Partnership, Coates Lorilleux Ltd, AMR Technologies, Malvern Instruments and Nanoforce Technology Ltd.

References

1. J. A. Darr and M. Poliakoff, *Chem. Rev.*, 1999, **99**, 495-541.
2. T. Adschiri, K. Kanazawa and K. Arai, *J. Am. Ceram. Soc.*, 1992, **75**, 2615-2618.
3. P. Boldrin, A. K. Hebb, A. A. Chaudhry, L. Otley, B. Thiebaut, P. Bishop and J. A. Darr, *Ind. Eng. Chem. Res.*, 2006, **46**, 4830-4838.
4. A. A. Chaudhry, S. Haque, S. Kellici, P. Boldrin, I. Rehman, A. K. Fazal and J. A. Darr, *Chem. Commun.*, 2006, **21**, 2286-2288.
5. E. Lester, P. Blood, J. Denyer, D. Giddings, B. Azzopardi and M. Poliakoff, *J. Supercrit. Fluids*, 2006, **37**, 209-214.
6. A. Aimable, H. Muhr, C. Gentric, F. Bernard, F. Le Cras and D. Aymes, *Powder Technol.*, 2009, **190**, 99-106.
7. S. Koshizuka, N. Takano and Y. Oka, *Int. J. Heat Mass Transfer*, 1995, **38**, 3077-3084.
8. Z. Shang, Y. Yao and S. Chen, *Chem. Eng. Sci.*, 2008, **63**, 4150-4158.
9. Q. L. Wen and H. Y. Gu, *Ann. Nucl. Energy*, 2010, **37**, 1272-1280.
10. B. Zhang, J. Shan and J. Jiang, *Prog. Nucl. Energy*, 2010, **52**, 678-684.
11. C. Dang and E. Hihara, *Int. J. Refrig.*, 2004, **27**, 748-760.
12. Z. Du, W. Lin and A. Gu, *Journal of Supercritical Fluids*, 2010, **55**, 116-121.
13. P. X. Jiang, Y. Zhang and R. F. Shi, *Int. J. Heat Mass Transfer*, 2008, **51**, 3052-3056.
14. P. X. Jiang, C. R. Zhao, R. F. Shi, Y. Chen and W. Ambrosini, *Int. J. Heat Mass Transfer*, 2009, **52**, 4748-4756.
15. X. Cheng, B. Kuang and Y. H. Yang, *Nucl. Eng. Des.*, 2007, **237**, 240-252.
16. J. Licht, M. Anderson and M. Corradini, *Int. J. Heat Fluid Flow*, 2008, **29**, 156-166.
17. A. Cabanas and M. Poliakoff, *J. Mater. Chem.*, 2001, **11**, 1408-1416.
18. ANSYS Fluent Package, www.fluent.co.uk, 2010.
19. W. Wagner and A. Pruß, *J. Phys. Chem. Ref. Data*, 2002, **31**, 387-535.
20. C. Y. Ma, C. J. Tighe, R. I. Gruar, T. Mahmud, J. A. Darr and X. Z. Wang, *Journal of Supercritical Fluids*, 2011, **57**, 236-246.

NUMERICAL INVESTIGATION OF TURBULENT HEAT TRANSFER ENHANCEMENT IN HELICALLY COILED TUBES USING AL$_2$O$_3$ NANOFLUID

A.M. Elsayed, R.K. AL-Dadah, S. Mahmoud and A. Mahrous

School of Mechanical Engineering, University of Birmingham, B15 2TT, UK.
Email: R.K.AL-Dadah@bham.ac.uk

1   INTRODUCTION

Passive heat transfer enhancement techniques can improve compactness and thermal efficiency of heat exchangers. They are preferred due to their simplicity, longer operating life, and lower cost and power requirements compared to the active enhancement techniques. Although there are various methods of achieving passive heat transfer enhancement, they all depend on changing flow geometry or modifying the thermo physical properties of the base working fluid. Helical coils, additives to fluids, swirl flow devices, rough and extended surfaces are all passive enhancement techniques.[1] Helical coils have been shown to be effective in enhancing single phase heat transfer,[2] boiling heat transfer,[3,4] and condensation heat transfer.[5,6]

Nanoparticles improve the energy transport properties of the base fluid by increasing the effective thermal conductivity and heat capacity, which enhances the heat transfer rate of the nanofluid. The chaotic movement of ultra fine particles accelerates the thermal dispersion process in the fluid which leads to a steeper temperature gradient between the fluid and the wall augmenting heat transfer rate.[7] The applications using these nanofluids include engine cooling to reduce the engine weight and fuel consumption,[8] increasing the critical heat flux in boilers[9] and developing compact heat exchangers for medical applications.[10]

Recently, many researchers experimentally[7, 11] and numerically[12-14] investigated the effect of nanofluids in enhancing the heat transfer in turbulent flow regime in straight channels. Particularly, the use of Computational Fluid Dynamics (CFD) techniques proved to be effective in simulating the fluid flow and heat transfer behaviour of nanofluids.[13,14]

For nanofluids flow in helically coiled tubes, there is limited experimental and numerical published work. Wallace[15] measured the heat transfer rate using nanofluids in helically coiled cooler however the author did not report any measurements of heat transfer coefficients or wall temperatures. Therefore this paper presents a CFD modelling study to investigate the heat transfer enhancement in turbulent flow with AL$_2$O$_3$ nanofluids through helically coiled tubes.

2 METHOD AND RESULTS

2.1 Flow Governing Equations and Thermophysical Properties

Al_2O_3 nanofluid has been treated as incompressible, steady state, homogeneous and Newtonian fluid with negligible effect of viscous heating. The flow has been modelled using Navier-Stokes equations solved by Fluent 6.3 CFD package. The single phase homogeneous flow governing equations in the Cartesian co-ordinates using k-ε turbulent model has been proposed for the current study.[13] The effective thermo-physical properties of the nanofluid are described by Elsayed *et al.,*[16] where the properties were formulated as user defined functions (UDF) subroutines and incorporated into Fluent 6.3 solver.

2.2 Heat Transfer in Straight Tubes

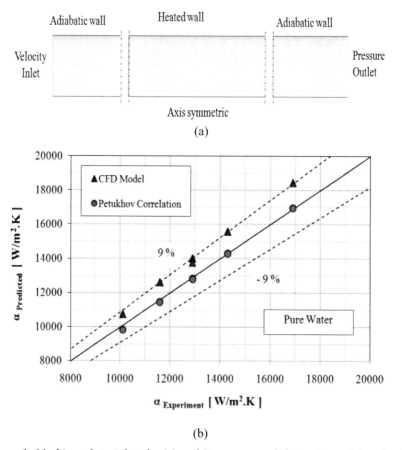

(a)

(b)

Figure 1 *Meshing of straight tube (a) and Pure water validation in straight tube (b)*

Figure 1a shows the boundary conditions and mesh configuration for a straight tube with 9.4 mm internal diameter and 2819 mm long.[17] Two adiabatic sections with 1 m and 0.5 m long respectively were positioned before and after the heated section. The heated section was meshed with 40 and 1600 nodes in the radial and axial direction respectively. The 1 m and 0.5 m adiabatic sections were meshed with 40x800 and 40x 400 nodes in the radial and axial directions. Second order upwind scheme was utilized for discretizing the energy and momentum equations, turbulence kinetic energy and turbulence dissipation rate. Uniform heat flux was applied to the heated section with uniform velocity at inlet to the 1 m adiabatic straight tube. The coupled algorithm was used with Courant number set to one for solving the pressure-velocity coupling.[18] The average heat transfer coefficient was calculated using the average heated wall temperature and average fluid temperature in the heated tube. Figure 1b shows the predicted heat transfer coefficient of the base fluid (water) and those reported by Williams *et al.*[17] with ±9 % agreement with experimental data and those predicted by Petukhov[19] correlation given in equation 1.

$$Nu = \frac{(f/8)\,\mathrm{Re}\,\mathrm{Pr}}{1.07 + 12.7(f/8)^{0.5}(\mathrm{Pr}^{(2/3)} - 1)} \quad \text{where } f = (1.82\log_{10}(\mathrm{Re}) - 1.64)^{-2} \quad (1)$$

Figure 2 presents the predicted heat transfer coefficient of Al_2O_3 nanofluid in straight tube compared to the experimental results of Williams *et al.*[17] at volume concentration ratios of 0.9%, 1.8 % and 3.6% and Reynolds numbers ranging from 8000 to 60,000 with ±12% agreement. Pak and Cho correlation[20] was in a good agreement with the CFD prediction. On the other hand, Vajjha *et al.* correlation[21] tends to under predict the experimental measurement and Maiga correlation[22] was found to over predict the experimental results.

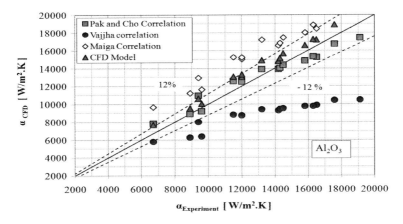

Figure 2 *Validation of Al_2O_3 in straight tube*

The developed CFD model was then used to investigate the effect of nanofluid volume fraction on the heat transfer enhancement ratio in straight tubes at various Reynolds numbers with 30 kW/m² heat flux. In this analysis, the heat transfer enhancement ratio is defined as the ratio of heat transfer coefficient of the nanofluid to that of the base fluid at the same inlet Reynolds number. Figure 3 shows that the heat transfer enhancement ratio increases with the increase in nanofluid volume fraction. The enhancement was close to 40% for concentrations of 3%. The maximum deviation between the Pak and Cho correlation[20] and CFD prediction was less than 7%. The effect of Reynolds number was found to be insignificant which agrees with most experimental measurements in the turbulent flow regime.[7, 23]

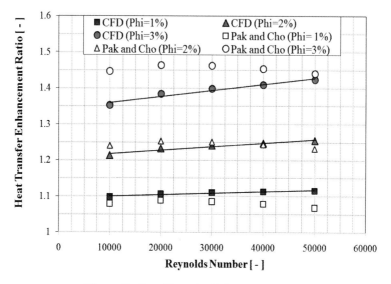

Figure 3 *Heat Transfer Enhancement ratio*

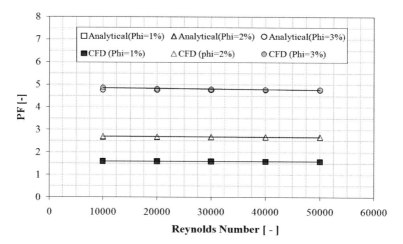

Figure 4 *Pressure drop penalty factor*

It has been shown that the friction factor of nanofluids agrees with that predicted by conventional theory.[7] Therefore the ratio of pressure drop for nanofluid and base fluid in straight tube for constant tube length, tube diameter and Reynolds number is expressed as:

$$\frac{\Delta p_{nf,ST}}{\Delta p_{bf,ST}} = \frac{\left(\mu^2/\rho\right)_{nf}}{\left(\mu^2/\rho\right)_{bf}} \qquad \text{where} \quad \Delta p = \frac{fL}{d_i}\frac{G^2}{2\rho} = \frac{fL}{d_i^3}\frac{\mu^2}{2\rho}\mathrm{Re}^2 \qquad (2)$$

Figure 4 shows the pressure drop ratio (PF) for the same Reynolds numbers and volume concentrations. The Reynolds number has insignificant effect on the pressure drop ratio. On the other hand, increasing the volume fraction leads to higher pressure drop ratio due to the increase in the nanofluids viscosity. Figure 4 also shows a close agreement between the CFD and the analytical predictions using Equation (2).

2.3 Heat Transfer in Helical Coils

A helical coil with coil length and tube diameter similar to those used in the straight tube (9.4 mm internal diameter and 2819 mm long) with 1 m and 0.5 m adiabatic sections has been modelled. The coil pitch was selected as 15 mm and number of turns of 5 leading to a coil diameter of 179.5 mm. The discritization schemes utilized were second order for energy, first order for momentum and SIMPLEC algorithm with skewness factor of one for coupling the velocity and pressure. The mesh contains 1,026,000 cell elements where the number of nodes in the axial direction were 500, 1500, and 250 for the inlet straight, helically coiled, outlet straight tubes respectively. Figure 5 shows the mesh used where Tri-quad meshing has been utilized to mesh the inlet face and hex/wedge cooper mesh used to mesh the coil volume with 6 layers close to the wall. The mesh quality has been checked by revising the turbulent wall function y+ value (less than 5 as depicted in (Figure 6)) and comparison to pure water empirical correlations. The required simulation time for each case was 8 hours using 2.4 GHz core Quad processor with 2 GB RAM memory computer. Figure 7 shows close agreement between the CFD predicted based fluid heat transfer coefficients with 30 kW/m^2 heat flux and empirical correlations of Seban and Mclaughlin[24] and Mori and Nakayam.[25] The absolute mean relative deviation between the CFD prediction and those of the Seban and Mclaughlin[24] correlation was found to be less than ±3.2 %.

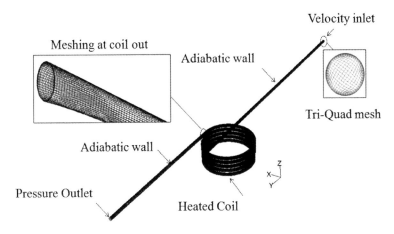

Figure 5 *Helical coil Meshing*

Figure 8 shows the heat transfer enhancement ratio versus the flow Reynolds number. Here the enhancement ratio is defined as the CFD predicted heat transfer coefficient of water in helical coils compared to that of water flow inside straight tube with the same diameter and length. It is clear from this figure that the enhancement ratio ranges from 1.07 to 1.12 which is considerably lower than those reported for the laminar flow of 2 to 3.[2,16] Also, the heat transfer enhancement ratio increases with Reynolds number in agreement with the findings of Kumar *et al.*[26] and Naphon.[27] Figure 9 shows the heat transfer enhancement ratio (heat transfer coefficient of nanofluid in the helical coil divided by the heat transfer coefficient of the base fluid in the straight tube with the same internal diameter and length) versus Reynolds numbers at various nanofluid volume fraction. It is clear from this figure that for Reynolds number larger than 20,000, the heat transfer enhancement ratio increases with both Reynolds number and volume fractions.

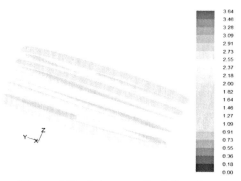

Figure 6 *Turbulence wall y+ function*

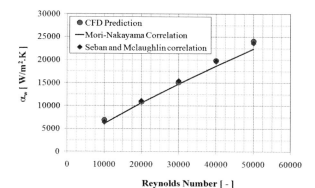

Figure 7 *Water validation in helical coils*

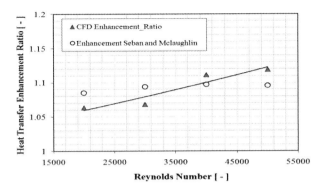

Figure 8 *Water heat transfer Enhancement in coils*

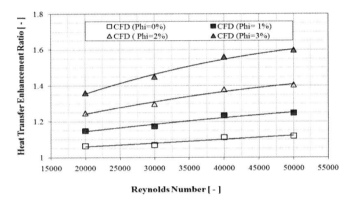

Figure 9 *Al₂O₃heat transfer Enhancement in coils*

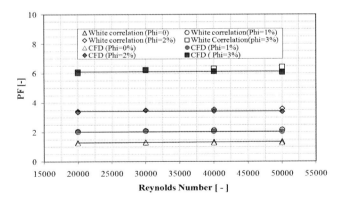

Figure 10 *Pressure drop penalty factor in helical coils*

The effect of nanoadditives on heat transfer in helical coils was found to be close to that of nanoadditives in straight tubes in the turbulent flow regime. However, the pressure drop penalty was found to be larger in the case of helical coils as depicted in (Figure 10).

The pressure drop in helical coils using Al_2O_3 for volume fraction larger than 2% exceeds 4 times that of water in straight tubes. For the same tube length and Reynolds number, the pressure drop ratio of nanofluid flow in helical coil to the base fluid in straight tube can be expressed as:

$$\frac{\Delta p_{nf,Hc}}{\Delta p_{bf,St}} = \left(\frac{f_{nf,Hc}L_{Hc} + f_{nf,St}(L_{tube} - L_{Hc})}{f_{bf,St}L_{tube}} \right) \left(\frac{\mu_{nf}}{\mu_{bf}} \right)^2 \left(\frac{d_{nf}}{d_{bf}} \right)^{-3} \left(\frac{\rho_{nf}}{\rho_{bf}} \right)^{-1} \qquad (3)$$

Where L_{tube} and L_{Hc} are the total straight tube length including the adiabatic parts and the coil length with 4319 mm and 2819 mm respectively. The friction factor of nanofluid in helical coil $f_{nf,Hc}$ was calculated using White correlation[28] for turbulent flow while the friction factor of nanofluid in the straight tube $f_{nf,St}$ was taken as equal to that of the base fluid in straight tube (Blasius correlation[29]) at the same Reynolds number.[7] Thus:

$$\frac{f_{nf,Hc}}{f_{bf,St}} = \frac{f_{nf,Hc}}{f_{nf,St}} = \frac{4\left(0.08\,\text{Re}_{nf}^{-0.25} + 0.012\,(d_i/d_{coil})^{0.5}\right)}{0.316\,\text{Re}_{nf}^{-0.25}} \qquad 15{,}000 < \text{Re} > 100{,}000 \quad (4)$$

Figure 10 shows that the CFD predicted pressure drop ratios are in close agreement with those calculated by equation 3.

3. CONCLUSIONS

Various heat transfer enhancement strategies in the turbulent flow regime have been investigated numerically including nanofluids in straight tubes, base fluids in helical coils and nanofluids in helical coils using Computational Fluid Dynamics (CFD) techniques with ANSYS – Fluent 6.3. CFD results were validated against published experimental data and correlations. Using 3% volume fraction of Al_2O_3 nanofluids in straight tube's turbulent flow has enhanced the heat transfer coefficient by 45 %. Also, CFD results have shown that 3% particle concentration of Al_2O_3 dispersed in water increases the heat transfer coefficient in helical coils by up to 60% of that of pure water in straight tubes at same Reynolds number. From these results it is clear that the helical effect in enhancing heat transfer coefficient in turbulent flow (only 10 % on average), is very poor compared to that reported in laminar flow regime (2 to 3.5 that of straight tubes). While CFD predicted pressure drop values were within less than ± 5% deviation from published correlations, the pressure drop in helical coils using Al_2O_3 for volume fraction larger than 2% exceeded 4 times that of water in straight tubes.

Nomenclature

C specific heat, J/kg.K
d_i tube diameter, m
d_{coil} helical coil diameter, m
Nu Nusselt Number, $\alpha di /\lambda$
Pr Prandtl Number, $C\mu /\lambda$
Re Reynolds number, $\rho u d /\mu$

ΔP pressure drop, Pa
λ thermal conductivity, W/m.K
μ fluid viscosity Pa.s,
ρ fluid density, kg/m^3
α heat transfer coefficient, W/m^{2}K
u velocity, m/s

References

1   A. E. Bergles. *Exp. Therm Fluid Sci.*, 2002, 26, 335.
2   V. Kumar, S. Saini, M. Sharma and K.D.P. Nigam. *Chem. Eng. Sci.*, 2006, **61**, 4403.
3   S. Wongwise and M. Polsongkram. *Int. J. Heat Mass Transfer*, 2006a, **49(3-4)**, 658.
4   A. Elsayed, R. Al-dadah, S. Mahmoud and L. Soo. Microfluidics, France, 2010.
5   S. Wongwises and M. Polsongkram,. *Int. J. Heat Mass Transfer*, 2006b, **49(23-24)**, 4386.
6   L. Shao, J.t. Han, G.p. Su, J.h. Pan. J. Hydrodyn. Ser. B, 2007, **19(6)**, 677.
7   Q. Li and Y. Xuan. *IJNA*, Science in China, Series E, 2002, **45(4)**, 408.
8   S.K. Saripella, J. L. Routbort, W. Yu, D.M. France and Rizwan-Uddin. http://papers.sae.org/2007-01-2141.
9   L. Cheng. Recent Patents on Engineering, 2009, **3**, 1.
10 L.S. Sundar, S. Ramanathan, K.V. Sharma, and P. Sekhar Babu. *IJNA*, 2007, **1(2)**, 35.
11 S. Torii. International Symposium on EcoTopia Science, Japan, 2007, 352.
12 C. T. Nguyen, G. Roy, P.-R. Lajoie and S. E. B. Maiga. 3rd IASME/WSEAS Int. Conf. on heat transfer, Greece, 2005, 160.
13 M. Rostamani, S.F. Hosseinizadeh, M. Gorji, and J.M. Khodadadi. *Int. Commun. Heat Mass Transfer*, 2010, **37**, 1426.
14 V. Bianco, O. Manca and S. Nardini. *Int. J. Therm. Sci.*, 2011, **50**, 341.
15 K. G. Wallace. Msc in Technology Purdue University, 2010.
16 A. Elsayed, , R. Al-dadah, S. Mahmoud, A. Mahrous. *Third MNF*. Greece, 2011.
17 W. Williams, J. Buongiorno and L. Hu. *J. Heat Transfer,* 2008, **130(4)**.
18 F. J. Kelecy. *ANSYS Advantage,* 2008, **2 (2)**, 49- 51.
19 A. Bejan,and A. D. Kraus. J. Wiley, 2003.
20 B. Pak and Y.I. Cho. *J. Heat Transfer*, 1998, **11**, 151.
21 R.S. Vajjha, D.K. Das, and D.P. Kulkarni. *Int. J. Heat Mass Transfer*, 2010, **53**, 4607.
22 S.E.B. Maiga, C.T. Nguyen, N. Galanis, G. Roy, T. Mare, M. Coqueux. *IJNMH*, 2006, **16(3)**, 275.
23 G.P. Celata. Nanotec2008, Venezia
24 R. A. Seban and E. F Mclaughlin. *Int. Heat Mass Transfer*, 1963, **6**, 387.
25 Y. Mori, W. Nakayama. *Int. J. Heat Mass Transfer*, 1967, **10**, 681.
26 V. Kumar, B. Faizee, M. Mridha and K.D.P. Nigam. *Chem. Eng. Process,* 2008, **47**, 2287.
27 P. Naphon. Int. Commun. Heat Mass Transfer. 2011, **38**, 69.
28 J. Welti-Chanes, J. Vélez-Ruiz and J.V. Barbosa-Cánovas. CRC Press, 2003.
29 S. Kakaç and H. Liu, 2nd edn, CRC Press, 2002, Chapter 4.

Subject Index